the best of
ROCK & ICE
an anthology

Happy 29th Birthday, Stu!

I thought this might be interesting reading for you. If you ...t to exchange it, there is a c receipt inside the book.

love you so much!! M,

the best of
ROCK & ICE
an anthology

edited by
Dougald MacDonald

THE
MOUNTAINEERS

 Published by
The Mountaineers
1001 SW Klickitat Way, Suite 201
Seattle, WA 98134

© 1999 by *Rock & Ice*

All rights reserved

First edition, 1999

"Paying for the Summit" excerpted from *In the Shadow of Denali* (New York: The Lyons Press, 1998), reproduced by permission. "Of Odds and Angels" adapted from *Fall of the Phantom Lord* by Andrew Todhunter. Copyright © 1998 by Andrew Todhunter. Used by permission of Doubleday, a division of Random House, Inc.

No part of this book may be reproduced in any form, or by any electronic, mechanical, or other means, without permission in writing from the publisher.

Published simultaneously in Great Britain by Cordee, 3a DeMontfort Street, Leicester, England, LE1 7HD

Manufactured in Canada

Cover and book design by Ani Rucki
Book layout by Alice C. Merrill

Cover photograph: *Trevor Messiah on Airy Interlude, The Needles, California.* © Andy Selters
Interior photographs: Page 6: © *Mark Twight.* Page 9: © *Jay Anderson.* Page 77: © *William M. Wood.* Page 139: © *Nick Papa collection*

"Lucille Has Messed My Mind Up" from the album "Joe's Garage," written and produced by Frank Zappa. © 1970 Munchkin Music. All rights reserved. Used by permission of The Zappa Family Trust.

Library of Congress Cataloging-in-Publication Data
CIP information on file at the Library of Congress

♻ Printed on recycled paper

contents

Preface *by Dougald MacDonald* 7

CLIMBS

Over the Edge *by Martin Atkinson* 10
Take the A-Strain *by Mike Bearzi* 13
Lucille *by Jay Anderson* 20
The House of Pain *by Mark Twight* 26
A Winter's Ghost *by Joe Josephson* 35
The Big Muddy *by Dougald MacDonald* 42
French Fried *by Ruaridh Pringle* 49
Playing God on Denali *by Michael G. Loso* 58
New York Stories *by Josh Lowell* 68

CLIMBERS

Penned in Leavenworth *by Alison Osius* 78
Climb to Safety (In Case of Flash Flood) *by Pat Ament* 84
Confessions of a Climbing Instructor *by John Long* 92
Largo's Apprenticeship *by Jim Bridwell* 102
Bad Boy *by Cameron M. Burns* 106
Verve *by Will Gadd* 110
Paying for the Summit *by Jonathan Waterman* 122
Prophet or Heretic? *by Eric Perlman* 129

OFF THE WALL

Credibility Gap *by Greg Child* 140
Revenge *by Jeff Long* 152
Monkey on His Rack *by Nick Papa* 161
The Zombie Traverse *by John Burbidge* 167
A Day Alone *by Barry Blanchard* 178
Night and Day *by Christian Beckwith* 184
To Die For *by Coral Bowman* 192
Of Odds and Angels *by Andrew Todhunter* 198

preface

Good writing about climbing is precious and rare. Seeking that "gem" among the hundreds of submissions that flow into the offices of Rock & Ice magazine is like waiting for a dream ice climb to come into condition. It can be a frustrating business.

When such a rare and excellent story makes its way into our hands, the background chatter in the office fades, palms begin to sweat, and that scene on a storm-bound ledge in the Alps or a tower of ice in the Rockies emerges from the page with consuming vividness. Fresh, original, riveting, these are the stories we rush to share with one another. There is a thrill in this discovery. As editors of climbing literature, we are often the very first to read a piece of writing that one day may meld into the collective consciousness of the climbing community. It can be a satisfying business.

What ranks the stories in this collection as "the best" climbing tales in Rock & Ice's fifteen-year history? We've identified two attributes: unique characters and originality. These stories represent the breadth of the climbing world, from bouldering to mountaineering on snowy peaks, and everything in between, throughout the far reaches of the earth. Yet, despite the variety, each has at its center a distinctive, memorable character or two, whether real or fictional, whether depicted in the third person or autobiographically. Each has a unique voice, from the quiet contemplation of Pat Ament to the swagger of John Long.

Rock & Ice has published nearly every top climbing writer, many in the early days of his or her climbing and writing career—Greg Child, Jeff Long, John Long, and Mark Twight among them. Today we continue to add strong new voices, combining youthful passion with the eloquence and wisdom of the more established climbers.

It's refreshing to see that the interests of our talented young writers span the range of the sport. At one end of the spectrum, among the great peaks is Christian Beckwith, editor of the *American Alpine Journal* and founder of the *Mountain Yodel*. During two trips to the snowy ranges of the former Soviet Union, Christian developed numerous friendships in Kyrgyzstan and the other mountainous republics, and he came to feel that the climbers there were as compelling as the rock and ice faces above their homes. "Night and Day" (1998) distills Christian's many hours of conversation and contemplation into a moving story about the struggles of these impoverished mountaineers.

At the other end of the climbing scale is Josh Lowell, whose "New York Stories" (1999) skillfully entwines character, language, and a sense of place among the tiny boulders of greater New York. Lowell, an expert boulderer and competition climber, is beginning to answer the often-asked question: Is it possible to write well about sport climbing and bouldering, where the old literary standbys of epic suffering and adventure have little meaning? Lowell's portrait of the enthusiastic New York boulderers, whose volcanic energy spews forth in barely decipherable English, offers a resounding "yes."

Rock & Ice has always been a home for the gifted amateur writer, and sometimes the previously unpublished ones rise to the top with the best known. Climbers like Jay Anderson, who returned to Wyoming for nearly a decade for repeated attempts on an off-width chimney called "Lucille" (1991), cause us to wonder at the power of human motivation. We are inspired by the craft of writers like Mike Loso, who was so struck by a near tragedy on Denali that he spent months investigating one poorly prepared climber's attempt on the peak. His exciting account of this climb—"Playing God on Denali" (1998)—is also a well-reasoned defense of the Denali rangers' policy. And then there are the truly bizarre stories, like the adventures of Nick Papa and his pet monkey, Fred—"Monkey on His Rack" (1994).

Rock & Ice was started fifteen years ago, when Neal Kaptain saw the need for an alternative climbing magazine in North America. The first issue had twenty-four pages, all black and white, with stories by Pat Ament, John Harlin, and Alex Lowe. Kaptain was succeeded shortly thereafter by George Bracksieck, who guided the magazine for thirteen years, armed with a vast store of climbing knowledge and a never-wavering taste for the surprising and unusual. Sally Moser, Nancy Prichard, Will Gadd, Marjorie McCloy, and DeAnne Musolf-Crouch are among the editors who helped tame George's vision. Under new ownership since 1998, *Rock & Ice* now comes out eight times a year and reaches more than 100,000 climbers around the world. Its editorial team—Julie Garrison, Aaron Gulley, Stephen Miller, Andy Moore, and Clyde Soles, under the art direction of Connie Poole—is the best this magazine has ever seen.

Not only do the stories in this volume represent several of the best days in the history of our editorial office, but they represent another new direction for the magazine. This book is the first of a series to be published by The Mountaineers Books, compiling the best information from the pages of *R&I*, as well as new stories by leading climbing authors. We hope you enjoy the old twists and the new turns as we make our way along this untrodden path.

Dougald MacDonald
Editor, Rock & Ice
Boulder, Colorado
May 1999

climbs

over the edge

Martin Atkinson
May 1988

With their rigid ethic against hammered pitons or bolts, England's gritstone crags have a special place in world climbing. And Master's Edge is perhaps the archetype of these climbs: a striking, square-cut arête, with exceptionally difficult climbing and a dangerous symmetry—there is only one protection point, exactly halfway up the climb. Martin Atkinson perfectly captures the tension of the third ascent.

The idea of having a certain route on a certain cliff is surely an abstract one, an arbitrary test left by a first ascensionist saying, "I have climbed here, can you?" This is undeniably the case with the latter-day eliminate. Yet some latter-day routes are rather more than just a statement of physical difficulty and inherent danger. What is it that bestows upon these routes a tangible presence and character? I don't know, but the Master's Edge will always be my example of such a route.

The sculpting of the Edge, which is striking even amongst the considered uniqueness of Millstone, is the work of one accomplished in the art of Black Comedy; it is serious and thrilling, yet paradoxically humorous. Between the ground and the jug which signals the end of the route's difficulties, the Edge sports but three useful holds (aside from the edge of the arête): a foothold at ¼ height, the shotholes at ⅔ height and a fingerhold at ¾ height. Their distribution is enticing; yet the distance between them is somewhat greater than the average man measures tip to toe.

The protection is assigned provocatively, placed precisely in the region of half height. The quarryman was kind, but not rash, to grant these holes. It is, of course, possibly coincidental that the two singularly most difficult moves are: (a) to reach the holes (at 22 feet, to jump from these moves is conceivable—to fall from them isn't); and (b) to lunge to the final jug which, at 18 feet above the runner at 22 feet, leaves little tolerance for rope stretch.

Perhaps the most perverse feature of the Edge is the plethora of undercuts littering its right wall. Of what possible use can a score or more of these be as you layback the arête facing the opposite direction? The assumption that these were deliberately placed where impossible to use lingers in your mind as you attempt to finger each and every one. This action, of course, only serves to interrupt the rhythmical flow of balance.

Hands and feet, each pushing, pulling, pinching and gripping, working sometimes with and sometimes against first this foot and then that hand—which in turn, sometimes pushes, pulls, pinches or grips—the whole concept is a little too difficult to rationalize. Yet, to succeed, the climber must understand an equilibrium and then instinctively sum all the incremental shifts of forces required to conserve balance as one limb moves six inches up the Edge.

So much for the subconscious. What is tangible is the experience, and it is this with which the conscious mind must play.

It is with the motivation of desire that one must approach a respected challenge such as the Edge, and, until this desire is fired, one is better employed elsewhere. One day, and for no discernible reason, I wanted to climb it.

To keep quiet is to avoid commitment, but since I have to borrow the only camming device that will work in the holes, this declaration of my intent is officially noted.

As a precursor to any significant event, there is usually a lesser, but in some way related, event. I remember once climbing a new route on the Gwynt. In the morning, for fun, we jumped off the lip of Pigeons Cave (40 feet to the sea); in the afternoon I fell 50 from the route. For my first attempt on the Edge, we journey to Millstone via a first-ever driving lesson. We arrive intact, and this seems significant. Several attempts to gain the queryman's holes find me unable to sort out the sequence. The retreat is orderly, via three foam pads and a second driving lesson. Tomorrow shall see.

Along with my increasing determination, the jump from below the hole grows harder to take. My knees compound into my chest and my neck is sore—whiplash?

With a respect for the jump born through experience, I have become sufficiently cowardly to risk falling from the moves by "going for it." The balance insured by the foothold is forsaken for paltry smears as both hands grasp the failing arête . . .

Amigo bottom right. Tri-cam top right. This at least has been practiced; in holes at the bottom of Xanadu they had refused to come out, giving a level of assurance now difficult to recall.

Gear, protection, safety—a situation with which I was familiar. The next ten feet of arête speed by in (I'm sure it was) one move.

Where am I? On an arête sans orientation. No top, no bottom, no holds, nothing solid to grasp. I suddenly see the runners and instinctively dive for their comfort; they pass me by, but the realization has returned and I float to the ground at the leisurely pace indicative of those proficient with a belay device.

Oh shit! I did it! There's gear in there, and better, it holds. I spare a thought for Wolfgang and wonder how he survived when his didn't!

The party scene masters the Edge and we all smile. A trot up the ledges to the right affords the view of a small fingerhold level with the last dab on the arête. "No problem—six inches up and left from the 16th undercut."

The trip to the holes is a giveaway. My Amigo points from its perch to the elusive fingerhold three feet above. I pirouette with my toe in this hole and take the section above in two-two time shuffle, hand-hand-foot-foot, repeat. I shimmy right to the fingerhold; balance is here and finger strength is the word. The party image visits me, but is lost as I leave the fingerhold to be alone, teetering in the breeze, eyeing the jug.

The wind gusts, surely I needn't say right to left, and I head toward the point beyond which there is no hope. But wait—somewhere approaching this point the wind falls off and leaves a little nothing to pull me back to where I was before. Capitalizing upon this good fortune, I launch at the jug—and miss.

The scream lasts good and long. From far above, I spy the holes containing my salvation. Far below, the rope becomes tight some feet from the floor as, without hesitating, Mark [my belayer] sets off up the Edge. I come to rest upon the foam pads, cosseted head to toe from the rudeness of this terra very firma. The laughter echoes in the neighboring corner, which is called Green Death, and which suits my pallor. "But you can't fall off any higher than that!"

True!

The weather spits and threatens, but the eye of the storm hovers above us and for this time the wind is elsewhere. Once again I stare at the jug, watching my right hand trace that self-same arc. In my mind's eye I see the fall again, but as I open my eyes I'm hanging from the hold—feet dangling and shins grazed from the Edge as my soles rip from the smears. First blood to Master and we are all square.

take the a-strain

Mike Bearzi
January 1989

Mike Bearzi is among the legions of unsung climbers who perform near the highest standards of their day but never become famous (usually by their own choice). He was putting up the hardest mixed rock and ice climbs in Colorado long before the current fashion for mixed climbing. Bearzi's account of Andromeda Strain, a classic Canadian Rockies alpine climb, captures the balance of outward bravado and quaking internal fear that characterizes many such ascents.

Tenuously, I drag my right foot from under the roof onto two small knobs covered with a light skiff of snow less than 12 inches above the lip. Form and technique are irrelevant considerations now. The foot begins to quiver and shake, yet I am strangely detached. The knobs move, then tumble off, the foot sliding and scraping down after them. At the lip, it lodges on a small edge. I freeze for an instant, hoping it will work. I've done this move too many times now.

"Cut! Oh, just beautiful, fantastic! Did you see that? Mike, let's do it one more time. Tommy, put those two rocks back and sprinkle on some more snow. Tom, God damn it, hurry! We're losing the sun! Mel, you bastard, you didn't tell me there were more clouds! OK, positions, tape, action!"

At this moment I can't help but reflect that exactly 72 hours earlier (it was actually 96, but 72 sounds better) I was grappling with the crux of the Andromeda Strain in the Canadian Rockies. Things felt pretty real then. Now I am grappling with a waist-high boulder at 14,000 feet on Longs Peak in Colorado, playing a turn-of-the-century mountaineer for a TV ad. The most real thing about this shoot has been the searing cold. I'm colder here than I've been in years. Hurry up and wait, stand around on pins and needles; feet numb, nose dripping. When the director says "Jump!" by God, you jump—no time to whip off a parka and stash it.

■ ■ ■

No climb is just a series of moves and pitches. There's background. There's history. Usually it starts with weird shit from your childhood, but I'll skip that. Eighteen months should suffice.

Eric Winkelman and I had not climbed together since I belayed his lead to the summit of Cerro Torre at dusk in a rising storm. Upon returning home, the piper demanded payment for this success. Our psyches depleted, there remained precious few assets to cash in. After a year and a half of struggle with this debt, true passion for climbing having become elusive and sometimes only a memory, the time came to take an accounting.

At first intent on a big new hard route on [name withheld], we settle on the [name withheld] face as a consolation prize—fun, but not the mirror we are looking for. The fact that our first project is out of shape sticks in my throat. This had been my story more than once, and I could hear it already. "Shit, Bearzi, are all of your routes out of shape? Sounds like you've sucked Winkelman into your act." After a brutal walk out, we forthwith make haste to the Columbia Icefields.

A half-hearted attempt on the North Face of Alberta ends in a long traverse along the base of the upper headwall and five rappels. Then, as if we have nose rings, we obediently trudge up two hours of moraine to glass the Strain. We're hardly speaking to each other. My legs are dead tired. I feel resentful. I don't care if that over-achieving cretin who is my partner wants to climb tomorrow, I'm not doing it. No siree, I'm too tired. At our vantage point, I catch up to him, suck in my belly, and say, "Eric, I'm wasted. I can't do it tomorr . . . "

Voice quaking, he answers, "Oh, I'm so glad you said that. Me too. I'm fried," or words to that effect.

Clouds part. A burden is lifted. We're pals again. Tomorrow we'll do breakfast in Jasper, then just be slugs for a day. In this brighter light, the Strain looks cool, real cool. We're psyched. We want it.

Eric, ever the linguist, dubs the crampon technique of frontpointing with one foot and Frenching with the other as "pied à splitova." So now we're in Splitovaville, the lower couloir, wandering along together as rappel anchors pass by at 150-foot intervals. Stimulating bulges brighten our lives now and then as we head along the base of the rock band to the highest alcove of the lower couloir. By doing so, we've thrown a major kink into an otherwise compellingly straight line, diverging some 40 feet from the main Strain drain. Well, that's what the topo says to do. Now the little dictator says to continue even farther right to a ledge on which, I suppose, one walks for 175 feet back to the true line. Our combined sense of purity piqued, we decide to take a ledgy shortcut back left. This will not only eliminate a pitch, but also will avoid 100 feet of a really disgusting chimney.

Eric is out of sight around a corner, moving slowly but steadily. I am slightly chilled, not particularly enjoying myself, starting again to wrestle with my self-doubt demons, and thinking maybe this sport ain't my bag any longer. Then I hear those dreaded words, "Hey Mike, this looks cruiser." Damn! Why did he have to say that? It's going to get desperate now. I stub out my cigarette and scrutinize my anchor. In a few minutes, the inevitable next words float around the corner and I silently mimic them. "Uh, Mike, this looks hard. Watch me."

"Gotcha, Eric."

Sometime later, I'm on the loose, balancing on weird ledges—not all that easy and the crux still looms above. A gap between ledges with only space below looks totally cruiser, so I steel myself for some moments of quiet desperation. I see evidence of Eric having rooted around in the rotten rock for some pro. From an arm bridge of sorts over the void, I remove a Friend buried in crud, leaving myself open to a rope-slashing pendulum. My first instinct is to sleaze my way through these moves. You see, by now I've decided that I don't want this route to be all that hard, and when we're done I'll quit alpinism, thereby copping to reality. Futilely, and probably pathetically, I cast about for some sort of tool hooks. All slopers. With a sudden rush of determination, I bare my hands and commit. Whoa, dude! This is dicey—frontpoints on quarter-inch holds, hands on rotten slopers. Gritting my teeth, I actually do it. The rest of the pitch is somewhat vague in my memory, because I am consumed with the happy thought that maybe I am yet a climber after all. A spot of thin mixed has a focusing effect and I'm with Eric.

We're back on that dotted line of purity. With a hint of nascent enthusiasm, I embark on this, the first of two crux pitches. I bridge and worm my way up a rotten chimney. Twenty feet above Eric, a small promontory that I use as a hiphold comes unglued but doesn't fall. Eric is a sitting duck, so I carefully pluck several cowpie-like discs and chuck them off. Aping up to some big slopers, I bridge across them.

"Hey, are you going to take my picture or what? Oh, and could you double shoot, please?"

The second click somehow sends my left foot skating off its hold. I'm left hanging by my left hand, my right knee basically in my mouth, a possibly fair-weather Friend eight feet below. Dumb! Concentrate, you idiot. A few more feet and I'm at a 20-foot A1 crack. It certainly looks free-climbable, but hey, the topo says A1. Who am I to argue? I happily motor it, then make more worm moves around a chockstone in this chimney-cum-corner. I arrive at a swath of thin yellow ice.

Gazing above, I size up the rest of the pitch. I don't mean to, but when I open my mouth to say, "It looks really, really, really hard," the words,

"Must've done the crux: this looks easy," come out. And then, just to drive the nails in the coffin, I say, "This route's in the bag, Eric." God, and I honestly believe it.

As a result of this delusional state, I'm feeling unusually frisky, which is okay, because things get hot and heavy right away. I find myself climbing in a schizophrenic style. My left shoulder and arm are performing very crude moves, mainly braced back against the dihedral wall, nudging up bit by bit, while my feet and right hand are involved in some very honed activities out on the face. Initially, I simply marvel at this. It's actually quite clever. It solves problems. But my friskiness proves to be quite thin indeed, and I soon realize, to my horror, that I'm being *dragged* up this pitch. I plead. I beseech. "Hey man, don't do that shit! You're scaring me! An adze/fist stack? You're crazy!"

As unwilling passenger, I play Mr. Stick-in-the-Mud and fret about the protection. It warrants caution. But there is something engaging about these guys, a certain earnestness that I find refreshing. I can't help but be caught up in the spirit. After tossing out a suggestion or two, I roll up my sleeves and join in wholeheartedly. By the time we reach the belay, it seems such a pity to stop.

As spokesman, I query Eric as to how much rope remains. The expected response would have been, "Not much," meaning, "I'm sick of hanging out here." However, "Sixty feet at least," wafts back up the dihedral. Now a finely tuned climbing machine, Team Bearzi is out of there. I've assumed leadership. I'm in the driver's seat now, directing operations, decisively barking out commands.

"OK, everybody out on the face! Back in the corner! Pinch that icicle! Fuck the topo; it sucks out there!" Et cetera.

Leeringly, a mushroom blocks the way, a thin seep of ice oozing from its belly. We ease up. Then I nag. I coax. I coerce. Tap, tap. Easy does it. Dance the feet up. Throw that bridge! Shaft plant, shaft plant! Now just stand up, dickweed, and you've done it. Standing up, I bang my head. It's a cave. Wow, what am I doing way up here?

After two hours, Eric has apparently reached his limit, because I'm informed that I'm out of rope. Suddenly, I'm overcome by strong territorial urgings. I stomp around, take a leak, and claim this hard-won turf. Fifteen feet of thin vertical will have to wait for the touch of Eric's tools before we can enter the upper couloir.

The ring of well-driven steel echoes across the Icefields, then the flick of a Bic and I'm ready to bring up Eric. This is one of my favorite parts of the climb. Eric starts raving about my lead. He inquires as to where I got a certain couple of body parts to do it. (This partially in Spanish, mind you.) Well, I was born with them, thank you, even if they have been retracted for a couple of years. Having freed up the aid crack, he keeps motoring along, all the time telling me he's gripped. I'm not sure I really believe him, but as I always say, "Flattery will get you everywhere."

The gear switch on top of the mushroom is cramped. I'm probably a poor host; besides my sense of property, I'm claustrophobic and hustle him off. In his precise and powerful style, Eric fires the last 15 feet and sets a belay just above.

I relish my first tool placement in the upper couloir. With my feet still on icy nothings, my tool, at long last, sinks to the hilt—sort of a visceral sensation.

You know, alpinism is supposed to be a man's world where one does manly deeds, thinks manly thoughts and strikes manly stances on north walls. Well, this can be a tough image to maintain consistently because, on occasion, something stirs from within, compelling you to shed this cloak, unclench your teeth, relax that steely glare, and be real. Unsuspecting, I grab the rack of screws and take off—Mr. Macho. Gray Bubbly ice. I let my tool fall of its own weight . . . socker. Crampons stick like I'm climbing a gigantic cabbage. I'm dancing up on front points. Splitova right. Splitova left. Anything I damn well please. Then it hits me. I'm happy. Just flat out happy. Happier than I've been in years, as a matter of fact. Not because of my lead, and not because we're going to bag this route. No, simply that some clot has kicked loose inside. It's like I can breathe deep again, or that a southern storm has blown itself out, or that I've quit beating my head against a wall, or maybe not a wall but a tenacious invisible membrane encompassing my life, only now there's no barrier anyway, and for the moment, no duality.

This is all kind of incongruous with the circumstances of the moment. Eric wants to know what I see around the bend. Manly voice, "Look's like it will go, Eric." Giggle. Who gives a shit? Isn't life wonderful? This would all be well and good if we lived in a perfect universe where all is symmetrical, because this little purgative episode of mine would have taken place at the top of the route and I could tie up this story with a neat little square knot and have a tidy package. However, the Strain isn't over. Above the upper couloir is the upper headwall, and above the upper headwall are the upper slopes, and then it's over.

Gray Bubbly gives way to Blue Plate. Happy feet give way to cramped calves. A 500-foot hyperbolic sweep of ice becomes a bowling alley. And I'm on a roll. We're simul-climbing and everything I knock down caroms off the side walls, sporadically and unpredictably nailing Eric. There's no place to hide. I'm sweating bullets, moving up on funky placements to somehow limit the ice fall. The topography of a double fall line creates a situation where I climb for 20 feet to surmount a single bulge.

"Ice, Eric." I say it loud enough for him to hear, but I don't, I can't yell it. It sickens me. I watch a big one begin its insane trajectory. Almost impassively, Eric tracks it, then flattens against the ice. Dead center. Now I yell. "Jesus! Are you okay?"

"Yeah."

He's not freaked, or isn't showing it, I don't know which. Moot point, though, because my mind has become twisted by the 150-foot runouts coupled with the real possibility of me knocking him off. His equanimity soothes my fraying nerves. The image of a rope and two climbers falling, whiplashing, twisting, then flying, crashing, mangled, begins to fade. I continue and, except for that one crater, I leave a string of huge, fractured plates *in situ* along this swell.

Bowling alleys don't go on forever, and, finally, our little scared rope is moving from its lower end. Eric arrives. He's bruised and has a coagulated drip of blood on the bridge of his nose. It looks kind of manly.

A pink sling dangling from a pin prompts a revelation. Eric and I have a debate as to what this means. I can't figure it as anything other than a legacy of someone's retreat—a hell of a long one at that. I feel for them. So maybe, I think, Andromeda Strain is more than just a route on Mount Andromeda that strains people. Maybe it's also a large strainer with three screens, each progressively finer, the last one placed diabolically high on the route. But I'm not losing any sleep over this. I've got my secret weapon—Eric. My attitude when it's his lead is, "This one's as good as done."

He deserved this lead. I hogged both crux pitches, but this one, besides being very hard and, by the looks of it, spooky, is everything a mixed pitch should be—assuredly exotic, but above all, photogenic. It would be terribly egocentric to say this is where the aforementioned main Strain drain ends; it may be the end as far as us climbing *up* is concerned, but stepping back and looking at it in terms of geologic time and the retreat of the Pleistocene glaciers and all, this is actually where it starts going *down*.

To exit this confusing situation, one traverses right to snag a dribble from above.

I lean back on the anchors to watch the show. A thin, sloping, angling, ice-crusted foot ledge cuts across a just off-vertical limestone wall of small rectangular blocks. Eric steps off the icy security of the couloir and into this concentration of adjectives. Pinching the blocks, he performs delicate step-throughs, being quite fussy about his handholds, throwing away any he doesn't like. Midway, he clips the pink sling. Twelve more feet and he sinks one tooth of his pick into the dribble, shifts a bit, and, using smear tactics, finesses his right crampon, unholsters his left tool, sinks it a tooth, shifts, imbeds three teeth of each tool, steps through, then "Socko!"—that sweet vibration of a well-placed tool. He drills a screw. We chat. He wants to rest and psyche up to run out the 30 feet of vertical above.

Eric climbs up the side of a big salmon draped over the headwall. The reflected light of the setting sun lights up its icy scales. No wait, he's scaling this fish. Oh God, how clever. How fun it is to climb a route of double and triple entendres. Quite carping, Eric. This is one whale of a pitch. Don't flounder now, big fella. Holy mackerel! He's done it! And so he has. I wrestle with one about how he's "fin"-ished, but it just doesn't work. Anyway, I have to go climbing. Eric reels in the big sucker. The pitch is hairball.

As much as you, the reader, would like to see this tale come to an end in the near future, we, the climbers, are thinking similar thoughts about this route. Ever eager to please, I hasten away from the belay. I feel pressured. Eric wants off, I want off. We still can't see the top. I split-ova my little heart out, round a bend, and see a monster cornice, lord of all it surveys. Judging by its size, I estimate some 42 pitches will get us out. There's no way off but up, I grimly recite. I put my nose to the grindstone, my shoulder to the wheel, resigned to the fact that this will all go unread.

Bonk! Somewhere late in the second rope length, climbing head bowed in despair, I run into something. Oh, look at the cute little cornice. Where did this little dickens come from? It dawns on me. Shortly thereafter, the Andromeda Strain dispassionately coughs up two specks of phlegm.

Two climbers stand on a lonely summit. A pale golden light suffuses the air. Crests and troughs roll towards a distant horizon of peaks finely etched on a sky of cool, glowing pastels. As the first evening star gently pierces the celestial fabric, there is not a breath of wind, only a fine permeating vibration. And if we're ever so quiet, so still, we can hear within this vibration the sublime harmony of 1,000 violins playing elevator music and . . .

"Mike! Mike, you silly boy, you. We've *got* to get going. It's almost dark."

"Oh, sorry. Say, do you know the way down?"

lucille

Jay Anderson
July 1991

Obsession is a driving force among climbers, and it surfaces in the oddest places. What else could draw a man year after year for humiliating and painful attempts on a single climb on the cold, high plains of southeastern Wyoming? Particularly when the climb is an offwidth chimney. Jay Anderson's Lucille is no longer considered the hardest chimney climb in the world. But his story of obsession with one route endures.

In Vedauwoo I found the ultimate wide crack.

It seemed like the place to look. The words Vedauwoo and offwidth go together like coffee and climbing. Even some of the face climbs there have token offwidth sections.

It's not true that all the climbs at Vedauwoo are wide and mean. Dogmatic wide-crack avoiders see the large fissures that lurk there and imagine *Jaws*. Scenarios become rumors, rumors become stories, and the tale they tell is of nasty five-inch cracks with sharp teeth and caustic venom.

Most Vedauwoo climbers don't even like offwidth climbing. They're forced to do it to get up various routes. They don't search it out.

But I do.

Ever since I learned that you could get inside 'em (a back-to-the-womb thing), I've been afflicted with a gluttonous offwidth jones.

■ ■ ■

At a certain point I realized that, although Vedauwoo may have the most offwidths per acre, it didn't, until recently, have the hardest ones.

In the '80s, Bob Scarpelli upped the ante as far as Veda-wide climbs are concerned. His routes Squat, Pretzel Factor, Bad Girl's Dream, Muscle & Fitness and others represent probably the highest concentration of modern wide climbs in a single area. These, as well as some of the older, easier classics pioneered by Gary Issacs, John Garson, Doug Cairns, Layne Kopischka and others in the '70s, have made Vedauwoo a necessary destination for the aspiring offwidth hardperson.

But there are harder wide cracks in California, Arizona and Colorado: The Owl Roof and Paisano Overhang; Improbability Drive; and Animal

Magnetism. Still, there was one crack in Vedauwoo that I imagined would prove to be more difficult than any of those. . . .

■ ■ ■

Lucille has messed my mind up, but I still love her.
— Frank Zappa, from "Joe's Garage"

It was 1979 when I first saw Lucille—a magnificent 40-foot roof with a squeeze chimney running through it in the corner, where it meets a vertical wall. It had the bulging, smoothly rounded lines of a Henry Moore sculpture. I couldn't believe it—how could a line so beautiful have remained unclimbed?

I tried to imagine it—tunneling sideways through a chimney, then thrashing upward. Hard 5.10 or so, I guessed—easier if hidden holds turned up. Little did I know . . .

First Bill Roberts and I attempted the crack directly beneath Lucille. Even this got us into trouble. Our first try at the four-foot fist-crack roof was foiled when we had to do a lichenectomy on Bill's eye. That night we watched TV and drank beer. A commercial for a record collection gave us the name for the first pitch—"Best of the Blues."

The next day we were joined by Bob Scarpelli. Bill led the pitch, with Bob and I following. Being "innocent, ignorant or insecure," we underrated it at 10a (if inflammatory letters to international climbing magazines can be believed).

Finally, we were in the cave beneath the big roof. To say it was intimidating, especially in those days of EBs and tube chocks, is like saying El Cap goes up "for a ways." There we were, isolated in the bowels of a dark, dank, chilly belay cave, the uneven floor paved with vermin poop—and before us, the roof swept outward into the blazing sunlight. The crack flared downward like an elongated cross-section of an inverted funnel, threatening to disgorge would-be trespassers.

We worked on the roof for the remainder of the day. Over a period of several hours, each of us tried it many times. After the exhaustive effort of trying to tunnel sideways, we discovered that we could use a foot rail and do a sort of 5.10 "walk" out to near the edge. I finally made it most of the way out, with the psychological protection provided by tipped-out tubes.

That got me to the hard part—the point where you have to move up is where the puzzle begins. Your toes are on a hidden, sloping edge. Your shoulders are in a bombay chimney that starts at mid-chest and is offset from the foot rail by almost two feet. You lean back over the abyss. The chimney is so flaring that you have to hoist your body up into a horizontal orientation and get your lower leg into a place narrow enough to jam knee-to-heel. Yet, two feet higher, it's too narrow to turn your head. You either look back at your tube chock, rocking on its tips, or out through stone blinders, into the abyss.

In either case, you can't see inside the crack where your arms, legs, and body are trying to make unlikely jams. You have to grope blindly, aware only of how far you have to go. After a few feeble attempts, we gave up until we could devise a better protection system.

Late that summer my father died and I went to California. When I returned to Wyoming, it was winter and nobody was climbing cracks—wide or thin.

■ ■ ■

Any girl that looks that innocent just got to be called Lucille!
   —George Kennedy in *Cool Hand Luke* as recalled by Paul Piana
      after last call on 25-cent beer night in The Operating Room.

I spent the spring of 1980 in Yosemite riding earthquakes and big walls, and didn't get back into offwidth shape until fall. When I placed a bolt at the end of the foot rail it began to snow. I tried the moves a few times before lowering off. Winter came and I called it a year.

Eighty-one was like '80—with the exception that, when I finally got to the climb and clipped the bolt, it broke off in my hand. Remember the defective bolt episode in the late '70s and early '80s?

In '82, I moved 400 miles away to Utah. Commuting became impractical and I didn't return to Vedauwoo until the fall of '84. Mike (Fred) Freidreichs, Greg Waterman and I went out and tried the climb, this time armed with Big Camming Units.

The BCU's worked perfectly. I was able to safely fall more times than I really wanted to. Cold reality hit me in the face like an old diaper—the climb might never go.

"That's it. I'm tired of this damn thing. I don't ever want to see it again! I'm never coming back!"

A few months later I was in Laramie for a wedding. I visited Bob Scarpelli.

"Are you going back on that climb? Because if you're not, I want it."

"It's all yours, Bob," I replied.

■ ■ ■

It was three years before I returned to Wyoming. I climbed in Vedauwoo for three weeks before even thinking about the big roof. Even though I'd abandoned it, we'd named it. It had become known as Lucille, after B. B. King's guitar, continuing the blues motif started with Best of the Blues. Somehow word of the route had gotten out. People I'd never met before in Yosemite, Paradise Forks, Joshua Tree, even on far-flung desert gravel piles, were asking me how Lucille was going.

With all this commotion, we decided to give it another shot, just for laughs. We set out armed with tunes. We soloed up Walt's Wall, the blaster in Fred's pack sending out a sonic wail, no doubt infringing on some athletic slabbin' greenies' wilderness experience. For the nth time, Fred led Best of the Blues. With the box booming at the base of the crag, I made a few attempts and, at my high point, came within less than a body length of the summit.

This was real progress! It changed my whole perspective on the climb. Fred was still skeptical, but hopeful. Just then, Little Richard's memorexed voice wailed from below: "Lucille!"

"That's the first time I ever thought this thing could go," Fred admitted when we got down.

We opted to take a break and rest before the next try. We rapped down to the sun. (Did I mention that Lucille is always in the shade and always cold? Even on a 90-degree August afternoon?)

Unfortunately, we discovered some U. of Wyoming Norwegian exchange students having a multi-keg, generator-run stereo party. We stayed for "a beer," but after a few beer relays and keg-spout-sucking marathons, the afternoon was shot.

The next day was the last chance to try the route before I had to take off to Arizona. (I'd moved again.) When I tried the climb, the efforts of the previous day appeared to have created more lactic acid than I could push through. We'd also climbed pretty hard for the previous weeks with too few rest days (at least for an old guy like me). I got into the hard section and just hurt too bad. I needed everything and could muster nothing. Rats!

For over a year, an arm-barring wound on my left elbow would make it too painful to lean on the armrest of the car.

■ ■ ■

Nineteen-eighty-eight. This thing had clearly gone on way too long. Visions of Lucille invaded my dreams. I was dating events in my life relative to attempts on the climb.

I had business in Wyoming, but decided that my primary goal in Vedauwoo would be Lucille. I was completely invested—wanting nothing more than to do that climb. I talked to Fred, and he was also psyched. He wanted this epic over with as much as I did. He'd spent $150 on wide pro.

After a day of warm-ups, Fred and I went up to the Hatbox. I'm not sure if that's the day he led Best of the Blues blindfolded—or with one hand tied behind his back. I led through the roof and topped my previous high point, but still didn't make it.

Then Fred tried it (the first time in all these years). Ten years of climbing fierce offwidths had honed him more than he'd thought. He made some of the hard moves before being launched into space. All of a sudden this was something within his sphere. We decided to rest and do some easier climbs and return the next week.

We rested and worked on some training climbs—like the third ascent of Pretzel Factor and the second ascent of Muscle & Fitness (5.11—Bob, really?) in an effort to "think wide." The desire to conquer Lucille built.

■ ■ ■

When the bolt broke in '81, Will Gilmer, my comrade on that attempt, and I considered toproping. I wasn't completely sure why we didn't. I was frustrated at not being able to do it, but for some reason we held back.

Likewise, as this project dragged on into the more conservative era of Reagan, Thatcher and top-to-bottom climbing, somewhere along the line we realized that we could have saved a lot of time (years) by employing the hangdog rehearsal strategy, a ploy that by 1988 was hardly controversial.

But we didn't.

It wasn't that we felt such a distaste for that style as we had "in the old days." I'd started the climb on one style and it seemed important to finish it that way.

Another temptation had presented itself. When I almost got it—on the last few attempts—slimy lichen had caused falls. It seemed almost stupid not to wire-brush these on rappel; I've certainly done this on other climbs. In a war council with Fred, we decided against it. This climb had already turned into a nine-year epic. Since we'd already tried so hard, we figured we might as well persevere and go the full classic, yo-yo, ground-up, traditional style. We weren't making an effort to sway anyone else's point of view on how to climb. It was more that we were going to get the full value Lucille had to offer us. It could at least be a lasting footnote to a passing style and a tribute to the climbers who thought enough about style to climb that way.

I remember thinking, "Today it has to go. This is my third day on the route."

The third day on that particular trip that is. I didn't even know how many times I had tried the climb in the past nine years.

"It has to go today."

I had postponed my travel plans for a day and had a thousand miles of travel before me. It had to go. I couldn't be that close and leave without doing it.

Fred told me that the belay was ready. The 5.10 lieback seemed shaky in the cool morning air at 8000 feet. After ten quick feet, I started the 40-foot roof. I reclipped the #4 Camalot left from yesterday. (In '84 it was a #4 Friend, in 1979 an #11 Hex.) Then I squeezed through the first constriction, my feet below on the toe rail, my upper body jammed into a down-flaring bombay chimney. I rested and got my breathing under control before I continued out sideways, clipping the next two pieces. The position seemed like a rest, since maintaining it was easier than what I'd done and what was to follow. I clipped the last piece accessible from the dwindling toe rail, a six-inch Big Dude.

Lichen ground into my scalp and chalk filled my eyes. As I felt my strength ebbing, I started the first 5.12 sequence. I went for it and almost got it.

When Fred went up, I expected him to get it. I rationalized: "I'll be so glad to have it over with, I won't go psychotic from the thought that I worked on it all these years and then still didn't lead it first." Fred came close, but not close enough. Lucille squished him out into space in the middle of a particularly difficult and insecure sequence.

We took an hour to rest—stretching, meditating and previsualizing the next attempt.

As it turned out, I had previsualized all wrong, but hung on long enough to be at the top, screaming and crying, with Fred screaming at the belay, and Alobar the Dog barking at the base of the crag. After all those years, after all the changes in techniques and equipment, I finally knew where the hardest wide crack was. To my knowledge, it is the very first 5.13 squeeze chimney, one of the small number of 13s put up in traditional style.

When it was Fred's turn to follow, he got going and climbed the hardest squeeze chimney in the world in perfect form.

The next day, Alobar and I drove home to Arizona. We took the long way, east through Cheyenne before heading south, so that I could see Lucille one last time from I-80. Appropriately enough, just as we saw it, KTCL played "Stone Free" by Jimi Hendrix.

Play it, Lucille.

—B. B. King

the house of pain

Mark Twight
January 1993

Few climbing writers have inspired such vitriol as Mark Twight. His dark musings on his own inadequacies (and those of others around him) earned him the nickname Dr. Doom. Twight's light-and-fast alpine climbs, by contrast, have required the kind self-assurance enjoyed by very few, as with the two new routes in Chamonix described here. Perhaps only when engaged in the act of climbing itself can Twight leave his demons behind and live unburdened.

Daniel saw him squatting perpetually on some north wall, enduring the storms and terrors of the great faces, a contemptuous eye cocked at his malignant gods. 'You cannot starve me,' he would say to them, 'more than I've always starved—nor cause me greater pain than I've always suffered—nor make me any lonelier.' And there he would preside forever.
—Roger Hubank, *North Wall*

The north face of Les Droites, after a hot summer season and little autumn snowfall, is a disaster zone. What little ice remains is thin, black and hard as the deck of a battleship. The loose rocks that didn't shriek off the face during the summer are stubbornly frozen together, hibernating, waiting for the spring thaw and the next wave of summer alpinists. The sunless and bitter northwest face offers a hostile invitation to those who linger after the "season" is over. Lest word get around, this wall is seldom discussed by the few who know that a new route still waits to be climbed. Complacency and fear swirl together into a cocktail of inertia, which is then disguised by rationalization: that conditions are no good, that it will be an easier line with better ice, that the cable car is closed and the bars are still open. . . . No one goes near the wall. No one mentions it. Everyone hides his eyes and pretends that labeling the climb an impossibility makes it one, and makes it so for everyone.

But I really don't care what anyone thinks. I do what I do. I succeed. I fail. Sometimes I'm lazy. I forget to turn on my answering machine. I live and breathe along with my problems and my work and my self-inflicted pain. I live in France because the mountains are high and cold and beautiful, but sometimes they are not enough, and days arrive with an ugliness even I can't endure. If I stay in one place too long, I start seeing only the bad, the good glossed over and unrecognized. By the time October rolled in, I'd had it up to my neck with kissing people's faces in greeting and stores that close from noon 'til three and the snotty chauvinism that reassures the French that their past glory still counts for something. They were pissing me off with their trumpets and mountaineering narcissism and the *nouvelle cuisine*. I decided to get a bit of payback by poaching the route on Les Droites, even though one of the locals had told me about it in confidence. I'm no different than any other prejudiced bigot; sooner or later I can feel a little nationalism and xenophobia.

Barry Blanchard and Jim Scott were hanging around my house, waiting for another go at the Eiger. After two inspired attempts in less-than-cooperative autumn weather, Jim gave up the grail and returned to the USA. I had four days before my flight to Kathmandu, so Barry sharpened his tools and, with every beer we drank, took a little more hardware off the rack. Binge drinkers that we are, Barry and I had already waded through my supply of alcohol, accidentally broken the windshield of my wife's car, thrown up across the lawn in a weird male-bonding ritual, and with all that out of the way, agreed to go climbing—something we hadn't done together since 1988, when we gave our (insufficient) all to Nanga Parbat and Everest.

■ ■ ■

Our talents differ. Barry is a master of the Canadian Rockies' style of alpine masochism, with several truly sick first ascents to his credit. I am a convert to, and proselyte of, the fast-food alpinism flourishing in Chamonix, mechanized not only by cable-car access, but also by VHF radios, Walkmans, helicopter transport (if at all possible) and the modern alpine climber's lust for speed. Our differences go even deeper, because Barry's a reformed asshole and cynic, a man who has adapted to the human race and the social conventions inherent in it. Barry is considerate of other people. At least when he's not, he recognizes that fact and tries to do something about it. I am not, and I don't. I am willing to cut it all away in order to have *my* way, to live how I want and for only as long as I want.

I've used the knife on my country, my family and, finally—with no small amount of hesitation and suffering and fear—my marriage. It wasn't clean. It was not pretty. I killed a part of me when I did it in. I slapped convention and everyone who believed in us in the face—and I think about it every day, and sometimes in the night as well. Marriage is another proud institution I have soiled beneath my selfish feet. But confessing does not negate the crime, absolve the guilt or make it easier for anyone affected. I walked away as I have usually done, leaving the wreckage to work itself out. They all take it personally anyway, whether my actions had anything to do with them or not, so why not let them have their solutions without cluttering the issue? I resolve my problems my way—alone.

Barry's compassionate nature had him comforting my wife in her time of great pain and need, offering solace with a bit more than his strong arms and understanding. In turn, she held him and helped put him back together hard on the blistered heels of his own divorce. But all that came after Les Droites; we climbed "Richard Cranium Memorial" before life came between us, before our solid bond was ruptured and going to the mountains together became a delicate proposition. . . .

■ ■ ■

The cable car was indeed closed and I complained about having to walk the 5000 vertical feet to Argentiere Refuge. Barry admonished me for my attitude, saying that mechanization has soiled alpinism's purity. To be sure, machinery has taken the place of tortured physical effort in some instances; but just as the use of Friends and, later, bolts has allowed rock climbers to concentrate on just climbing, the system of high mountain refuges and competent rescue services has allowed alpinists the same luxury. Modern routes in the Alps reflect these resources.

Our rack was anorexic and we wasted time getting gear to fit. The potential for big, ugly falls became enormous, and a healthier, heavier rack would have let us move more quickly on the sparsely plastered rock; if either of us fell, our next stop would be inside a can of *foie gras*. With a Stopper and a knifeblade between himself and the belay, Barry broke his pick. He managed to place a nut, hang and replace the pick. Twelve feet higher he broke another one.

I sucked my balls up into my stomach as Barry swore, screamed and pounded the hammer against the wall. I thought he was going to jump and just get it over with. I empathized. Why bother climbing beyond capacities of the gear? I mean, we evolve and we get better, but it's too expensive for a company to change its molds or research new steel or plastic, so climbers are the ones who pay. I hadn't brought a spare pick or a third tool—breaking one of the rules I credit with keeping me alive up to now—and I promised myself not to forget next time. . . . I followed the pitch, grabbed the meager gear sling and cast off into the heights.

Barry reached my belay, breathing heavily, and told me I was climbing as well as Doyle does when he's fit. I liked hearing that, because Kevin is one of my climbing heroes.

"Mark, let's trade tools so I have two good ones to lead with . . . and maybe we could take the third anchor out to give me a little more gear to protect the pitch."

I agreed with his reasoning. "When I come up, Blanch, keep me tight so I can go faster."

"That's something we do in Canada, too," he said, arranging the gear on a sling. "Glad to see full-strength runners on your rack, Twight. It makes me warm all over."

"I'm even more paranoid about gear than before. I've just broken too much, or seen other people on the ground after they've broken theirs."

"Except for your ice tools, I notice. You don't have a spare pick, do you?" Barry's tone said he already knew the answer.

"Grivel designed these prototypes especially to be Mark-proof," I dismissed my oversight with a cavalier shrug worthy of any Frenchman.

We raced up the face, hunted by the coming night and nagged by our light packs, empty except for the walkie-talkie, headlamps and a virtually gasless stove—all of which would prove useless if we had to bivouac. But sleeping en route is very unfashionable in Chamonix. At 5 p.m. we were rock climbing about two pitches below the summit ridge.

I pulled through the last moves with my crampons screeching across the granite, thinking that nothing's ever a giveaway in this game, and then watched the sun disappear beneath the horizon as I belayed Barry up. Total blackness fell while we shared the remaining bits of food and frigid water. Barry started trundling rocks as he prepared a bivi site—sage tactics, as he had no idea how to get down off the south face of Les Droites (nice light show, too).

"Yo, bubba. What are you doing?"

"Aren't we going to sleep here? I mean, how're we gonna get off this pig in the dark?"

"Blanch, we're in Chamonix. We can't sleep here. The Couvercle Refuge is about 3500 feet down, with blankets and beds, and that's where we're going. Besides, I've got this descent totally rehearsed, with the rap anchors in place and everything ready. We're in the modern world."

I'd done the descent two years earlier with Philippe Mohr (who later died on the Aiguille Sans Nom), and during the trip down with Blanchard on that moonless October evening, I thought about Philippe and cried silently, wondering about our futures. We were snug and warm inside the Couvercle Refuge before midnight and back at my house packing for Nepal by 10:00 a.m. the next morning.

■ ■ ■

In all his life, he'd never known anything like this. Now he realized that the worst of life had always been like this. That always it must be like this for someone, somewhere. And perhaps the time had come for him to suffer what all men must have suffered since the beginning.
—Roger Hubank, *North Wall*

Barry Blanchard and Andy Parkin have little in common except for me, alpinism and *the painting*. Andy painted it, and Barry likes it. The painting hangs on my wall and Barry tells me that our lives are (more accurately, *my* life is) becoming more and more like it. A man's large, tortured and searching face first arrests the viewer's attention, and only very slowly do you notice the figures behind him. One is gesturing for him to come back (or pushing him away—it's hard to tell which). A naked woman lying on her stomach with her chin on her hands is eyeing him boredly, as if to say, "So long. Anyway, I've got these others." Another man looks on with a curious detachment as Mr. Tortured stares into his clouded future with his back to them all. The lines are hard and harsh and the painting is somber blue and gray and black. Barry casts me in the role of the tormented man and himself and my wife as two of the other figures. I like it.

Andy is, like Barry, one of my climbing partners. He was one of the world's best alpinists in the late seventies and early eighties, having soloed Les Droites by several lines, done the Walker Spur alone in winter (in 19 hours) and the serious Boivin-Vallançant on the Aiguille Sans Nom. Then, in Pakistan, he climbed Broad Peak and attempted K2 in alpine style. But a 1984 groundfall nearly killed him. It left his hip in 13 pieces, many of his organs displaced inside his torso, and his left arm shattered, along with his future. Today both the hip and the elbow are fused into single, motionless pieces. Despite doctors' predictions to the contrary, Andy *has* climbed since then—and climbed well. In the last three years, he has put up five big routes (all of which are unrepeated) in the Mont Blanc range, attempted Makalu and Everest, and climbed Shivling. He consistently climbs 5.11 and gets in the occasional waterfall routes when his career as a painter and sculptor allows him the time. Andy is one of the most gifted climbers on mixed terrain that I have ever had the pleasure to climb with. His determination, experience and willingness to risk it all propelled him up three new modern alpine routes in the Chamonix Aiguilles last winter. Beyond Good and Evil on the Aiguille des Pelerins is the most serious and difficult of these climbs.

The Aiguille des Pelerins' pitiless north face is an austere, monochromatic wall. Its shades of oppressive, life-threatening gray lighten as the sun passes over without touching; they never warm to red. Great men have left little of their presence on this face—merely three seldom-repeated routes. It is a cathartic place, attracting only those few who want to throw the dice, to beat their heads against it. All efforts undertaken in moderation count for nothing up there; every movement the heart does not believe in, every word the hand does not write in blood, is wasted effort, squandered without reason.

I was beaten on this wall. I spent my blood and my energy for nothing. I tried to build a monument to my efforts and I failed, leaving only some pins, tears and piss. The climb left a scar that time wouldn't heal, and I vowed to return and acquit myself. I could see the face from the gym where I train, and it tortured me to think that all the weights I'd lifted and the hills I'd sprinted up meant very little on that great north wall.

Andy and I had twice tried to climb a new route on the Aiguille des Pelerins, but were stopped once by technical difficulty and how slowly we progressed against it, and the second time by variations on the same and a turn of the weather. Our first attempt was in November 1989, when the days were criminally short and the belays interminable and cold. I took a 25-foot upside-down fall out of the big corner on the fifth pitch. We were demoralized by the difficulty and the relentless approach of the 14-hour night and unknown terrain waiting above us. We retreated from the top of the sixth pitch, turned our backs and snowshoed away.

For almost two years I wore the route around my neck like a weighted chain. Andy and I both had excuses for not going up again, the same excuses I criticize others for using: it never really comes into condition, the weather forecast is no good, our work is in the way and the cable car is still closed. We recited the usual list of rationalizations, substituting them for fear and laziness and lack of motivation. Neither of us were ready to give what we knew the wall was going to take. Finally, in April 1992, we clicked and meshed and started up again. It hadn't become any easier in the interim—conditions were not ideal and the aid went slowly. We managed to climb seven 60-meter pitches (the long rope was our latest secret weapon) to get a look at the upper wall before the clouds moved in and the snow forced us down.

With failure stuck sideways in my throat and sickness in my heart I wrote catalog copy (that didn't need to be written) for the following nine days until Andy called and said he was free again. I dropped everything and filed my ice tools into lethal instruments. For this, the real thing, we decided to get on the wall and stay on it until we finished. We took bivi gear and a pair of jumars so the second could follow and carry it. We loaded up the packs with metal and rope because experience dictated a rack of hardware larger than anything I had ever taken into the mountains (except for the South Pillar of Nuptse). The nine Camalots, 13 nuts, 13 pitons, two screws and I don't know how many carabiners were a huge judgment against our ability and confidence, but previous attempts proved this to be the minimum. I was quite embarrassed to be seen by my friends as I boarded the cable car with such a huge pack. It was a decidedly un-French affair, but if we succeeded (and then managed to sandbag some poor soul into trying to repeat the line), we could tell them all where to get off. I knew it was right for me, and I wasn't about to subscribe to the current "do-it-in-a-day" fashion, as it would only result in another washout.

The first pitch was brutal by headlamp, adrenaline on top of coffee and maybe the chance to see breakfast again. I knew I could get a good Lost Arrow about thirty feet up, and ten feet beneath it an iced-up crack took a #5 Camalot, which might have slowed me down.

"Spindrift!"

WHACK!

"Glad I wore the helmet. Hmm, the pick's buried about halfway and it only dragged a few inches before it caught on something. Too bad there's nothing for the other one." Looking down: "That Arrow doesn't look as good from above. Maybe I'll try wiggling a nut in there, yeah, and tap on it with the hammer. Now I'm hanging off the hammer. Well, I'll use the pick of the other tool. There, that shouldn't come out. Andy might have trouble with it, but that's his problem. He's comfortable down there with a three-piece equalized anchor and my down jacket on." Above, on easier ground, I decided that since I'd taken a long time below, I should run it out without stopping, "but that nut and the pin are way down there—can't even see 'em with my headlamp—oh hell, if I fall off this, I deserve the deck. It's only 100 feet to it, and the snow's deep and soft."

Jumars made following reasonable, since the passage of the leader often left no ice or snow for the second to climb. But the seventh pitch started with some unreasonable aid off copperheads and tied-off pins directly above the belay. Pulling out of the etriers into a 70° corner stuffed with just enough ice to mask the crack, I slowed down to face the music. The corner was capped with an overhang. Andy said that, on the first attempt, he'd found a place for an upside-down ¾-inch angle, but the latest storm had plastered a huge snow mushroom under the roof and I was afraid to touch it. (My imagination had me tapping it and then, as it broke loose, hugging it like a pillow as I sailed 30 airy feet, trying to keep the cushion between myself and the wall, hoping it might break my fall a little.) Instead I dry-tooled out left onto a slab with my right calf shaking uncontrollably and my frontpoints dancing a psychotic carnival step against the granite. I torqued the shaft of my tool in a wide, flaring crack and locked off low enough on it that I could hook a thin flake with the pick of my other tool. Gingerly weighting it, I leg-humped an arête and tried to rest. The only hope for gear was a slot for a wide Leeper; I hand-placed the pin halfway and beat the hell out of it, fixing it for the next generation. The pin gave me just enough confidence to reach a small shelf and good ice. "Resting" on the six-inch ledge, I drove in a long Bugaboo and equalized it to a nut, backed it up with a Camalot and my heart slowed down little by little.

Two exciting pitches higher I made the evening radio call to get the weather report while we dug out a bivi ledge. Everything was working in our favor; the bivouac was large enough for two to curl up and the stormy forecast had been modified to call for a few snowflakes caused by thunderstorms forming elsewhere. We slept as well as one does in these places and woke to broken, timorous clouds with the promise of amelioration during the morning.

After holstering his hammer following the tenth pitch, Andy heard a noise and turned quickly to see his hammer tumbling down the wall. He was forced to remove pins with the adze of a Barracuda. Placing them was worse (he couldn't get them in very far) and he fell 15 feet onto the belay as he started the 13th pitch—of course it had to be the 13th—where the belay was suspect. I had one hand in the pack searching for some food when his pin popped. Unwittingly, I gave him a dynamic belay, but it's probably what saved us from going to the ground.

He finished the pitch, and I followed it, warily eyeing the heinous slot clogged with big, loose blocks that gave access to the Col des Pelerins and the end of the climb. The only belay anchors were directly beneath this deadly cavity and it was too steep to avoid pulling on the flakes and teetering bricks. I was quite happy to be absorbed in the process of leading rather than waiting for the sky to fall on me. I fought with every ounce of my self-control that remained after 13 hair-raising pitches; I pulled gently, and pushed resolutely down and inward on the creaking mess. Tunneling through the cornice well above my last piece of protection, I felt the familiar greasy fear-sweat running freely from my armpits and the small of my back. It cooled rapidly as I belayed Andy up and relished the world's greatest remedy, victory, and my second nightfall route-finish of the year.

The snow was hideous, a slight crust over heavy, wet glop that often sucked us in to the waist. We rappelled and down-climbed the west side of the col and began the long march back to the cable car midstation where there was a small hut. My newly scarred watch showed 2:30 a.m.

when we reached its drifted-in doorway. We'd been on the mountain for about 45 hours and spent 26 of them actually climbing. My hands were smashed-up and bleeding, I was dehydrated and wasted as badly as I have ever been, and the knee I'd broken several years before was inflamed and pulsing dully. I swung gently at the end of my rope and had just enough energy to force the door open and collapse inside. I kept my pain to myself, though, because I felt it was certainly trivial in comparison to what Andy must have been going through. I admired his drive and commitment to the ideal.

Only great pain is, as the teacher of great suspicion, the ultimate liberator of the spirit . . . it is only great pain, that slow, protracted pain which takes its time and in which we are as it were burned with green wood, that compels us philosophers to descend into our ultimate depths and to put from us all trust, all that is good-hearted, palliated, gentle, average, wherein perhaps our humanity previously reposed. I doubt whether such pain "improves" us—but I do know it deepens us. . . .

—Friedrich Nietzsche

And we are chained together in the house of pain, searching for our truths—beyond good and evil.

a winter's ghost

Joe Josephson
January 1996

A single crag or canyon can occupy a central place in a climber's heart, drawing him or her back year after year. Yosemite is such a place for some. Patagonia for others. For Joe Josephson, an American who lived for years in Canada and wrote the ice-climbing guide to the Rockies, that place was the Ghost River Valley, a remote and trying setting for winter adventures.

For some people, looking at photos conjures up previous moods or experiences. For me, the strongest bond to the past is sound: the voice of a partner, the lapping of a lake, the stillness of high-mountain tundra on a calm evening. It is a cognitive zephyr that brings back images of people, places and emotions that move the soul. Thus, when I hear the heady Cajun song "Wild Injuns," I am reminded of the Ghost River Valley.

Located in the Front Ranges northwest of Calgary, the Ghost is always an adventure—and never the same. It is a place of disappearing rivers; lost souls of World War II POWs; offroadin' Alberta rednecks; miles of untouched rock walls; and steep waterfalls that lurk in the Valley's shadows.

For decades, the Ghost has been known to offer the best limestone in the Rockies. Yet it took until 1982 to discover its vast winter treasure. The first bite of Ghost River Valley ice came not to "real" climbers, but to local ranch hands wearing Wranglers and snowmobile suits and using rudimentary gear. Belaying from the bumper of their monster truck, they toproped one short icefall. Finding the experience only slightly more fun and far less glamorous than bull riding, they leaked word of their prize to some Calgary ice zealots. The next weekend, Iain Stewart-Patterson, Mike Blenkarn and Dave Bean went to vanquish what the cowboys couldn't. They rode to the bell with The Good, the Bad and the Ugly — an instant roadside classic. The remote corners of the Ghost began to be tapped, and major routes are still being found 15 years later.

This never-ending search for the Ghost is born from the feeling it breathes into you. It's a place to feel alive—alive as your legs and lungs burn on three-hour approaches; alive with the blossoming willows and grass that brush your boots en route to a February ice climb; alive as the wind rips your belayer's words across Devil's Gap, Bastion Wall, The Recital Hall, Valley of the Birds, Spirit Pillar, Phantom Crag, or any of the other mythical citadels in the Ghost.

The Ghost cast its magical spell on me in winter 1988, the first time I descended the infamous obstacle known simply as "The Big Hill." Gateway to the Ghost River Valley, The Big Hill is the steep, rocky decline you're obliged to torque, spin, bounce, push or pull your vehicle down—and up—to get in or out of the Valley. Fail, and you're left to contemplate an 18-mile walk. It's at The Big Hill that your adventure either begins or ends.

■ ■ ■

Big routes in the Ghost require a heady commitment to uncertainty and adventure. To feel up to this challenge, I needed several seasons under the guidance of Ghost River veterans—guys with big trucks who knew both the road and what lay beyond it.

By winter 1990, I had a four-wheel-drive Subaru and a sense of the Valley's challenges. My partner Jeff "the Naz" Nazarchuk and I were following the winding gravel road that precedes The Big Hill; we were en route to Waiparous Creek, a side drainage to the Ghost, and Hydrophobia, the biggest, baddest ice route the Valley has to offer. Unspoken apprehension filled our minds. What shape was The Big Hill in? Could we get through the drifts or rivers? Was the route formed? Naz had been to Hydrophobia, but he failed to even attempt the route after his partner took one look and declined the offer. I didn't know what the route looked like, but I knew its sustained nature, the four-hour approach and its history represented an intensity befitting the Ghost.

On the first attempt of Hydrophobia, in 1986, Karl Nagy and Frank Campbell barely escaped with their lives when the route released 10 minutes worth of its icy, barely liquid sustenance as they neared the top. Soaked to the bone, they struggled to the ground with ice-entombed gear and shorted headlamps, then stumbled home.

Few routes pull you in such a way that you court disaster, then return the following weekend. Hydrophobia has such a pull, and the duo returned to complete a climb of legendary proportions. The few subsequent parties have all been haunted by the possibility of a dousing by the great wave from above. These water releases are rare, I told myself. What's the chance of us going surfing? But hey, this is the Ghost.

My new Neville Brothers tape startled the cows as Naz and I raced toward The Big Hill. Time lapsed as we talked, laughed and carried on with a nervous exuberance that only the Ghost brings on.

The song "Wild Injuns" describes characters costumed for a Mardi

Gras celebration. We felt like Mardi Gras injuns, not necessarily pretty, but wild with the celebration of adventure and friendship.

The Subaru bounced down The Big Hill with aplomb, clearing the first test. We continued deeper into the Ghost, past rocks, drifts and rivers, and arrived at the trailhead. Our plan was to hoof it in with huge packs, beat a trail to the base and bivi. Two hours into the approach march, we passed a route even more famous than Hydrophobia. The Sorcerer preened, just like in the photos—a 600-foot, white-blue viper snaking up among Dolomitic towers. Naz had climbed the Sorcerer several years earlier, and his tale of traversing out of a comfortable cave onto technical, overhanging mushrooms on the last pitch filled me with intimidation. And Hydrophobia is harder?! In this corner of the Canadian Rockies, the last mountains between us and the Atlantic, lay some truly wild climbs.

Two hours later, in the soft, early evening light, our icy tower appeared at the back of the cirque. Reaching to the top of the cliff in one uninterrupted flow was a 400-foot plastered flow of frightening proportions.

It was a beautiful thing.

■ ■ ■

"Well, Naz, we've only a smidgen of fuel left," I reported the following morning. "Do we cook this soup, or do we light a fire?" Without hesitation, we made the biggest fire I've been ashamed to light.

So ended our first attempt at Hydrophobia. Our demise was not the free-flowing water that had stopped Karl and Frank, but rather free-flowing air. Arctic air. During the night, it poured in at minus 31 degrees Fahrenheit. We had overcome big hills, rocks, river crossings, car-eating drifts and bushwhacking, but the Ghost held the final trump. My Subaru, now dubbed "The Ghost Buster," scored its second coup by starting on the first punch, passing boot camp as only the Ghost can deliver it.

We were now fully committed to the route. Big ones don't fall easily in the Rockies, and we had to dispel the mystique of Hydrophobia for ourselves. One month and a spate of other climbs later, Naz and I once again found ourselves at the base of what we now called "the Phobia."

"This is the most desperate lead I've ever been on!" hollered Naz from halfway up the first pitch. In the belay cave, I blindly cursed him for taking so long and finding it so hard. But how soon my attitude adjusted. Had I ever seen such aerated ice? I couldn't even see my feet through the bulges and dollops of head cheese. Having the rope above my head allowed me to gently hook the lacy candlestick in a vain attempt to save strength in my undertrained arms. Next, stark fear entered my being as I struggled to remove Naz's Snarg. The ice piton marginally protected a 20-foot traverse Naz did to avoid even worse ice above.

By wedging my body between two fragile icicles, I was able to chop out the Snarg without barn-dooring into space. The overwhelming specter of a King Swing into oblivion kept me attached to the climb just long enough to get into good ice below the belay. "Nice lead," was all I could muster as I clutched his hanging stance. For the next two pitches, the angle remained 80 to 90 degrees. The ice only got better as we neared the lip, and as darkness spread across the eastern plain, we found ourselves on top of Hydrophobia.

■ ■ ■

Since then, Naz and I have climbed together on routes that are harder, easier, longer, shorter, on rock and on snow. Looking back, the memories that most fill me with life are of those days of discovery: searching for the mysteries of the Ghost, the limit of our abilities and the joy of friendship. The Phobia was pure adventure—not the longest climb, not the most epic, but for me the epitome. It's with buddies like Naz that I enjoy "pushing the envelope" and "cranking the rad"—not the popular-press rad, but our rad—wherever that may lie. When we found ourselves on top of Hydrophobia, we felt as though we had really accomplished something. Maybe we weren't the first or the fastest, but for us, at least, our climb was the best.

I was thinking these thoughts in 1992, as my now well-traveled Subaru lurched toward the Ghost in the pre-dawn ink. The tapedeck no longer worked, and there was no one to talk to except the cows—and they were asleep. I was heading back to Waiparous Creek....

Familiar landmarks pass by in the dawning Front Range sky. I'm a pilgrim in search of a new experience, perhaps a fresh way of seeing. Today I envision the Sorcerer and Hydrophobia together, alone.

Within the first few feet of starting Hydrophobia, the commitment sinks in. I'm pumped and wondering what's going wrong. In the previous months, I'd done many routes that were rated harder; none had made me feel this way. I'm a slave to my burning forearms as I struggle to maintain control. Confidence and joy of movement are replaced with pure survival.

Grading ice climbs has become simple for me. They are: Easy, Hard, I Can't Climb It, or I Won't Climb It. Today the Phobia is somewhere between Hard and I Can't Climb It, and it's quickly drifting toward the latter.

The concentration coefficient is immense. Every placement takes time and energy. The weather is mild, so why is the ice so brittle? My entire focus is on the square meter surrounding each of my ice tools. I have to be certain that, at any given time, one pick is into gold. As in that other great Canadian winter pastime that I've learned to respect, no hooking is allowed. Hockey's two minutes won't be the penalty here, but rather 100 feet to eternity.

The first half consists of short, overhanging steps separated by long stretches of vertical. Taking a needed break at the first possible stance, I see a previous party's rappel screw. I look at it dumbly. For some reason, backing off doesn't enter my mind. All I think of is the verticality ahead. It doesn't seem rational. Things that truly come from the gut rarely are.

I break it into segments; stretches of surmountable ice in front of me. Thus, the remainder of the climb seems to go much better—more rests appear, as does better ice, with only marginal periods of desperation. Nearing the top, I think of Naz's remark as he was leading in the same position two years earlier: "Oh man, I'm getting a biceps cramp. Watch me!"

It is my turn to be duped. It looks so good; just off-vertical. "My fave," I think to myself. The last 30 feet provide some of the best ice you could hope for. However, that is but a passing thought as the angle kicks to off-vertical—the wrong way.

The energy coursing through my system will not let me blow it now. I'm too close. I focus entirely on the horizon of the approaching lip. Then, to no excitement or fanfare, I stand on top of Hydrophobia, wonderfully alone. I wander upward, seeking the sun along the ridge. My mind is in a state I have never known. The concentration has wiped my thoughts clean—*tabula rasa*. Comprehension is beyond me as I meekly stare out into the foothills. I begin to calmly see myself from afar. I am a small speck in an improbable place surrounded by a landscape that makes me feel privileged. I smile at the thought of seeing myself as a tiny, colored fleck amongst the ice, rock and sky.

■ ■ ■

Doggedly, I make a spectacular ridge traverse to the top of another drainage several miles south of the Phobia. Here I entertain thoughts of my next objective—the Sorcerer. After some down-climbing, cornice-chopping and two rappels, I find myself in the large cave below the crux. I am wondering what the hell I'm doing. The peace I found on top of the Phobia an hour ago is wavering. I sit in the cave for a very long time. I had been at the same belay less than a month prior with a partner and opted out of the final pitch. "One of those days" "too cold" "burned out" "not into it" "I've done it before." I went through all the private excuses imaginable. Again, I can't figure out what motivates me. Why solo these routes in the middle of nowhere, four hours from anywhere, which is another hour's drive from somewhere? Does part of it lie within the routes themselves?

Sitting in the cave, the question "Why?" froths in my skull. What am I doing here? I am certainly not going to solo the Phobia again. "This is my only chance," I say out loud. It sounds so ludicrous, I laugh. Chance for what? To be the first to solo either of these routes, and to do both in a single day? Accomplishment? Notoriety? Death? Honestly, the Phobia provided some of the scariest 30 minutes of my life. How I long for some other party to be here. A polarizer to say I'm crazy and that it's okay not to do this, or, perhaps, to be an audience that adds an extra edge.

Carefully, I make the next rappel to a small snow slope. I contemplate going up from here. It would save me from climbing the easy bottom pitches. It would be easier on my still-aching arms. Yet it seems shallow, and I can't bring myself to do it. More rappels, and the bottom greets me. I rest again. Everything that led me to this point passes in front of my reeling imagination. I go full-circle and return to the vision and uncertain reasons that brought me to this special place. Without any explanation, I cautiously pick my way up the familiar ice, taking the easiest line to the right. Quickly and without thought, I pass by those previous points of indecision. Thick concentration returns as I continue through the overhanging mushrooms above the cave.

■ ■ ■

I need it. The fear, the risk, the ability to work through the laziness and uncertainty to reach my goals and tap the potential I see in myself. I needed it on Hydrophobia with Naz in 1990, and I need it here alone in 1992. These routes are my instrument. I haven't felt this way in months. All those climbs, all those pitches—they were fully predictable, no problem.

Soon there becomes no more fear, no more pump, just brutal immersion in wild movement up a seemingly hostile landscape. Without realizing it, I find myself cruising toward the top in a steep but restful groove. I start to laugh.

"I'm going to do this," I say, barely audible to myself. The giggling, the relief, feel enormously beautiful. What have I done? I have survived. I have realized my vision that these routes can be climbed in this way. Yet I keep coming back to the nagging thoughts. Where did this vision come from? And, more important, where did the drive to fulfill it come from?

Going out with partners like Naz is great—exciting, fun times with an important friend. There certainly was some ego involved with our original ascent of Hydrophobia; we were proud to have been two of the few to have done the route. But as history passes, it becomes just another ascent. Is it a need for ego gratification that pushes one to solo these routes? Does anyone really care?

My sense is that it is a coming home, like a small-town boy returning after college. He comes back with a breadth of experience, an inner knowledge and a spirited vision. That's what I feel in Waiparous Valley this day. The Phobia is a personal yardstick—it was in 1990, and it is again in 1992. I have returned from grander locations and harder struggles to a place I feel is home. It is here that I challenge myself.

To be here with a partner would be just another ascent. To solo any other route in any other drainage would be nothing more than just another energetic day out. Perhaps it would be impressive to others, yet I would find it shallow and ephemeral. This day, alone and in a personal arena, is a grand event. It is for me alone, and it will fade when I do.

Only the Ghost will remain.

the big muddy

Dougald MacDonald
March 1997

If you like first ascents, there's nothing quite like discovering an underexploited area, where impressive routes are just begging to be climbed. But if those routes are guarded by rotten rock, a frightening lack of protection and a three-hour four-wheel-drive approach, there just might be a reason why so many obvious lines are untouched.

I hate this road.

The truck grinds over another slickrock bulge and into a patch of cobblestones like relics from the North End of Boston. My head raps against the roof of the cab. It's my third time down this desolate track in less than a year. I thought the first time would be the worst, but then I broke my tailbone in a bike crash, and now I'm trying to stand up and keep my weight off my seat and somehow still shift, brake and pump the gas in rapid succession to keep my truck crawling down the 31 pounding miles to Monument Basin Overlook. I should give up the wheel, but I'm not yielding. At least it gives me something to hang onto.

Why do I keep coming back here? Unclimbed desert towers. I've got the itch, and Monument Basin is a good place to scratch it.

Monument Basin is a convoluted sandstone amphitheater about two miles in diameter and 400 feet deep. Within its overhanging walls are 20 or more free-standing towers and countless attractive fins and buttes.

Attractive, that is, if you know what you're in for.

Steve Komito didn't know. In an old article now legendary to desert rats, Komito recounted his fearful visit to Monument Basin with Layton Kor and Huntley Ingalls in 1962. Kor lusted after the first ascent of Standing Rock, a mighty, narrow-waisted cylinder more than 300 feet high. Shocked by the consistency of the tower's rock, Komito penned an immortal description: "The bottom of Standing Rock appears to be composed of layers of Rye Krisp held in place by bands of moistened kitty litter." Hyperbole, but the imagery has kept most climbers away for decades.

A coffee-table book on the Utah desert had clued me in to the potential. There was Standing Rock, right in the middle of a big photograph, but what were all those other towers? As far as most climbers knew, only two Monument Basin towers had been climbed: Standing Rock and the Shark's Fin, an outlandishly overhanging blade of stone that was featured on the cover of *Rock & Ice* 10 years ago.

I called Cameron Burns, a Colorado climber who knows a thing or two about desert towers. He said that, in fact, many of the spires in my photo had already been climbed, including the Mock Turtle, the Enigmatic Syringe, the Meemohive and the brilliantly named Staggering Rock, just over from Standing Rock.

Two men had bagged most of these towers in 1990 and 1991, together or solo: Roger "Strappo" Hughes and Steve "Crusher" Bartlett. Both men lived in Boulder. Both were English expats, weaned on rubbley rock in their homeland. Both had silly nicknames. Were these prerequisites? I hoped not. Many unclimbed towers remained.

Cam faxed a map showing the named formations. He had scrawled a note across the page: "Obviously, you're onto a good thing. As I'm a greedy, tower-hogging bastard, I'm not going to tell you any more."

Well, that sealed it.

■ ■ ■

I recruit Dave Goldstein for an exploratory foray. He has done Standing Rock, so he knows how to get into the Basin. This isn't as easy as it might seem. Six hours of highway, followed by hours of four-wheelin' on the rough White Rim Trail, and we're only to the Basin's edge; we still have to find a way past the rimrock to the floor of the huge bowl.

When he approached Standing Rock the first time, Dave scouted for an entire morning until he found a fixed anchor hidden in a sandstone crevasse. A short rappel led to a skidding scramble down steep sand and talus, with loose blocks tumbling around his ankles and knees. On the way out that night, he startled a scorpion on a key handhold.

When we finally reach the basin floor this time, we camp below Standing Rock and scout for new lines. The best is the biggest, a massive fin of rock between Standing Rock and the Shark's Fin. *Rock & Ice* #16 had jokingly captioned a photo of the latter formation as "Shark's Fin, Mars, where, because of feeble gravity and strong prevailing winds, rocks don't grow straight." Somehow, this name had made its way into the guidebook. Science fiction into fact—our unclimbed tower would be called Mars. Perfect.

The bottom of Monument Basin is a beautiful, lonely place. Except for an occasional shout from mountain bikers on the White Rim Trail, it is silent. Few tracks mark the soil—even lizards are scarce. A tiny herd of desert bighorn hides under the rim. The sky is empty but for an occasional raven or canyon wren, perhaps a pale half moon. The floor is sparsely vegetated and littered with chert, a colorful rock with a prosaic name.

The climbing is harder than it looks, and it doesn't look easy. Our line is a succession of bulges, like a fat man's stomach. These push into our chests, so the next aid placement always seems just out of reach. We scrape dried mud out of grooves, expect good cam placements, but uncover only slots that gape outward like half-open clams. The line is in the shade, and belaying is frigid duty. The November wind changes direction six times during the first pitch. A steady rain of sand and chunks of rock falls from above. I hide under an overhang, and I can tell Dave is getting higher when the chunks really begin to whistle. It takes three days to climb the seven-pitch route.

With one day left, we eye a fine virgin tower in the middle of the basin. Dave is already calling it Bruce Smith, which sounds like the name of a Mormon prophet but is actually that of a defensive end for the Buffalo Bills, a future Hall-of-Famer. Dave is a rabid Bills fan. If you look at the northwest side of the tower, you can vaguely make out a human torso, especially the leg and foot. Dave wants to climb the tower, take a picture of it and send the photo to Bruce Smith with a note saying, "You may have lost four Superbowls, but no other player has a desert tower named after him." It's this kind of plan that makes me at once eager and leery to climb with Dave.

Three days on Monument Basin's 60-grit stone have shredded our hands. We are taping from our forearms to the tips of our fingers. In this, if nothing else, each of us resembles Bruce Smith.

At the base, the tower is the usual appalling mix of loose flakes, stacked blocks and caked dirt in the cracks. But there is a difference. The cracks on this one nearly touch the ground, which suggests the possibility of free climbing.

Although it seems ridiculous at first, free climbing is often an option on Monument Basin towers. The underlying rock is solid, horizontal features abound, and pockets appear in blank rock. Crusher Bartlett says, "Some of that rock is really good, and the free climbing there is some of the most interesting in the desert. All these weird holds and pockets — not just a straight-up crack."

We climb Bruce's shin and his bad knee (a detached pillar), then find the 5.11 crux at an undercling around his butt. Above the fourth belay rises Bruce's burly neck and helmet, a slender, rounded capstone, like a giant baguette balanced on dirt. Forty feet of improbable face climbing over three bulges leads to a stance below the helmet. I drill a poor bolt, stand up and surrender. I'm sure I can get onto the baguette, but I'm unwilling to risk the downclimb. Dave takes over, mantels onto the top, and then, like a rodeo cowboy wrestling a calf to the ground, manages to swing back down.

At the rim, my truck's battery is dead. We have to wait a day before two four-wheelers come along and give us a jump start. I hate this road.

■ ■ ■

We craved unclimbed towers, and our desire had corrupted us during our first visit to Monument Basin. Canyonlands National Park has a new climbing-management plan banning all fixed hardware and piton placements. Yet, although we tried to climb hammerless, we had placed about five pins and seven bolts on Mars for anchors and protection. Bruce Smith took a lead bolt and two anchor bolts. Not much iron for 12 new pitches of desert climbing, but illegal all the same.

On the way out that time, a ranger had waved us over and asked what we were doing down in Monument Basin. Climbing, we said. What kind of climbing? A pause, and then a craven answer: Well, clean climbing, of course.

I hated that lie, and I vowed to play by the rules on my next trip—not just because of the rangers, but also because it would enhance the game. Only three of the ten towers that had been climbed in Monument Basin were known to go hammerless and, thus, were legal to climb. Clean ascents were a natural challenge.

Paul Gagner and I camp a mile from the truck on the south rim. Outside the basin, we see yet more unclimbed towers, including a slender needle below camp. I am excited about the possibilities, but it isn't the anticipation that keeps me from sleeping. At home, my marriage is rockier than the White Rim Trail, and it strikes me that I'm hiding from my problems in one of the most isolated places you can reach by car in the Lower 48. That night, I lie awake for hours in my bag and listen to Paul snore. At two in the morning, I step outside the tent into fresh snow and watch a comet glow overhead.

The next day, three friends join us to bag the first ascent of the Pixie Stick, the needle below camp. The summit is the smallest I've ever been on, consisting of stacked discs of rock, like sandstone pancakes. We cut a length of seven-millimeter static line, loop it twice around the biggest flapjack and rap off. It seems more degrading to leave 30 feet of rope on a desert summit than two bolts and a couple of slings, but rules are rules. The purple cord matches the rock nicely. Later the park superintendent tells me even this violates the rules. The standard is no new anchors, whatever they're made of.

Paul is getting into Monument Basin free climbing. "It's like mixed climbing, on rock and ice," he says. "You have to weight everything equally so it doesn't all rip."

The next day, Paul gets to test his skills on the second ascent of the Enigmatic Syringe, a tower that looks as if it were built by a kid playing with blocks. There's a big cube at the base and a cylinder balanced on top. Paul thinks the first pitch looks free-climbable.

A short corner and a traverse protected by a bolt lead him into a no-man's land of crumbling rock and bottoming seams. He painstakingly works out some protection and commits to a series of desperate, irreversible moves, with pro falling out and the possibility of a back-breaking fall into the corner below. Dave follows the pitch and declares it hard 5.11 and psycho. "The parts that looked bad are really bad, and the parts that didn't look bad are still bad," he exclaims.

I'm dozing in the sun, thinking I'll just jug to the summit, when Paul yells down, "Come up here and lead the second pitch!" I find the others on a huge belay platform. You could hold a square dance here, which certainly seems preferable to leading the summit column.

Strappo didn't waste any time placing good bolts on the first ascent. Angle pitons project preposterously from shallow holes, more than six feet apart, in the nearly blank rock. This pitch might go free to someone someday, but I'm not even going to try. I heel hook and pinch the Syringe's edge to reach between the aid pins, and lasso their eyes with a sling when I can't get to them otherwise. I think of Strappo soloing on the first ascent. If he blew it, his flesh would have been sandblasted off his bones before anyone even figured out where he was.

Paul wants to try the second or third free ascent of Standing Rock, but I'm now completely undone by thoughts of home. Tower climbing suddenly seems like a useless, selfish game. I sit on the rim and stare into the vast desert while Paul and Donna Raupp climb.

Paul blows his first try on the crux third pitch and lowers to a no-hands stance. The move is a deadpoint to uncertain holds, protected by a 34-year-old Star-dryvin. Even from the rim, I can hear him complaining that he has no chalk. I had proselytized against using chalk on desert towers, and, to his credit and regret, Paul listened. He sticks the move next try, though, and, when he gets back to the truck, he says the stone on Standing Rock is much better than that of the other towers in the basin. In another 34 years, when climbing is a major sport and the inevitable spin-off of people who prefer adventure to sport routes has ballooned the number of climbers in Monument Basin to perhaps 20 or 30 each year, the rock on the other towers may be clean too.

We have doubled the number of towers in the basin that go hammerless. But on the long drive out I can think only of the troubles that await me at home. In either direction I travel it, I hate this road.

■ ■ ■

Six months later, I am once again driving the White Rim Trail, trying to keep my butt in the air, seeking solace in my aiders. I have been seduced by the wild Northeast Arête on the Shark's Fin, first climbed by the prolific desert pioneers Art and Earl Wiggins and Katy Cassidy. This alluring line seems a fitting way to conclude my Monument Basin campaign. Dave and I had spent an hour scoping the line with binoculars and have convinced ourselves that it might go hammerless.

Others aren't so sure. Mike O'Donnell, who did the second of the route's two ascents, stacked three pins at the lip of a six-foot roof near the top. "No way will it go clean," he told Dave. Someone in our party had mentioned the plan to John Middendorf, who replied, not-too-cryptically, "Hmmm, that's a Wiggins Route, isn't it?"

The first couple of pitches are moderate, with only a few poor aid placements and some hairy free climbing. On pitch three, the rock tips to Rifle steepness and, 40 feet up, the cracks disappear. Hanging from a poor piece, Dave scouts for a solution for nearly an hour. The first two parties had nailed a thin seam to move across the fin. Dave tries all of our clean-aid techno-solutions but can't get anything to stick in the slim pin scars. He is despairing when he notices a rail of rock at his feet. An absurd Tricam between two teeth of sandstone and a couple of hook moves lead him across the fin, with a penalty slam waiting in the corner if anything goes wrong. Slightly better pieces lead upward again, but now it is nearly dark, so he ties about nine pieces together with all of his slings and his aiders and lowers off. Fifty feet up, fifteen feet overhanging.

The next day, in a single pitch, I am forced to place six different pieces I've never even used before. Seemingly bomber nuts settle half an inch or more before holding. When nothing appears solid enough for body weight, the moves go sort of French free, with one hand and foot on the rock and the others on the aiders. In the middle of the pitch's crux, I find a miraculous pocket off to the left and stuff a Friend and a Tricam into it. Following, Dave points out that I somehow neglected to clip into these life-savers before I climbed the 20 tenuous feet to the belay.

The crux is the big roof, and two fixed pins get Dave quickly to the lip. He spends the next 90 minutes dangling from his fifi hook and trying to place another piece. He tries all the conventional pro on the rack, then moves into the weird stuff. A hook placed upside-down in a pocket, tied off to reduce leverage, actually holds his weight for a moment but then blows out. The nut pick doesn't last that long. Dave's back is screaming from hanging at the lip for so long. Finally, he asks me to dig out the cheater stick and tape a nut to it. The first two he tries are too big, and it's the smallest nut on the rack that finally sticks, poking halfway out of a slot, four feet above the roof. A heel hook and a mantel, while pulling on this ridiculous placement, get Dave to easier ground.

On top, the original entry in the register reads "No Fun" and "Sorry Charlie, FA's are for hoseheads." I'm disappointed they felt that way. The Wiggins Route is by far the best climb of this type I've done—the steepest multi-pitch route in the desert.

All around us are unclimbed towers and walls. There's the needle we've dubbed "the Awl;" the 300-foot "Hidden Tower," nestled secretively among the fins of the south rim; and a bizarre cluster of towers that Crusher describes as the Mount Rushmore presidents with bags over their heads. Severely overhanging hand cracks split the White Rim caprock.

Some of these routes will be climbable under the current Canyonlands policy; some will not. If the regulations ever change, Monument Basin will produce some of the finest nailing routes in the desert. But the more climbers experiment with hammerless techniques, the more they find possible today.

"In a perverse way, the regulations do work," Crusher says. "They've forced me to think more about clean climbing or free climbing, and maybe they're pointing the way to the future. You don't have to nail if you can free past the unprotected sections." In 1995, Crusher and the talented British all-arounder Stevie Haston freed the Fetish Arête on the Shark's Fin, the lower-angled analog of the Wiggins Route.

A huge storm has moved in and the basin is riven by lightning as we head for home. In the pounding rain and darkness, I peer through my windshield trying to spot cairns that mark the way across the wet slickrock. Drop-offs loom alarmingly.

To pass the time, we speculate about the Shark's Fin. The Wiggins Route might go free to a real lunatic, someone willing to risk 5.12 moves on poor rock, 15 feet out from his last decent pro. I might even be tempted back to follow him on that climb.

Whoa, wait a minute! What am I thinking? I am never, ever, driving down this road again.

french fried

Ruaridh Pringle
March 1998

The epic is a staple of climbing literature, from The White Spider *to* Annapurna. *Avalanches, rockfall, cold and storms all take their toll in these tales, but surprisingly few climbing stories involve lightning, perhaps the most frightening force of all. In this story of storm and salvation, set among the Chamonix Alps, Scottish alpinist Ruaridh Pringle experiences a lifetime of terror in one climb.*

The corner is vertical and capped by a 45-degree roof. It is dark but for faint lines of illumination, so regular they could be man-made—granite edges catching light from somewhere. Their outline is familiar. I know this place.

Below my tiny ledge is a sensation of appalling void, somehow fascinating. Suddenly, I am off the ledge and falling, but then I find myself pushed gently back on, panting, insides scrabbling. Am I tied on? I peer about desperately in the gloom, searching for something. A way out. Nothing, even if I had a rope. Panic wells up. I'm stuck here. I'm stuck, oh shit!

I stagger to my feet and, after a few heart-stopping steps, find the light switch and offer thanks again to my anonymous benefactor. My bedroom. Not that infernal corner.

I examine my fingers. Not swollen, but still numb and scabby. Jeez, I'm dripping sweat. Legs still not right, but all considered I'm in better shape than I've any right to be. Not quite sure yet how I feel about that. The worst storm in 25 years. Central Chamonix a river of mud, and me strolling back to greet shocked friends as if I'd just popped out for dinner. Just a week ago. Strange how the nightmares came only in the last three days.

■ ■ ■

Hugo was 19, his first season in the European Alps. Huddled at the foot of the Mont Blanc massif, Les Chosalets campsite becomes a British climbing colony in summer, and it was there, in July 1996, that Hugo first approached my tent, with tales of broken promises and enormous blue eyes full of frustration. Had he sufficient facial hair, he could have shaved it in the sheen on his Camalots. I hadn't even seen a full set before.

"Which shop did you rob?"

"I said my rack was embarrassing. I got all this with a grant from my old school."

"Bloody hell. Does the big Camalot double as a parapente?"

After a week together pushing our grades in the granite playground of the Chamonix Aiguilles, we felt fit and confident. The Bonatti Pillar—the dominating feature of the soaring 12,238-foot Petit Dru—had tantalized us across the creeping ice of the Mer de Glace, but ambiguous forecasts and Hugo's lack of mountain experience had kept us away.

Immortalized by Walter Bonatti's legendary solo first ascent in 1955, the pillar was probably the most famous big rock climb in Europe. The Bonatti Pillar was freed in 1982 by avoiding the sensational aid crack of the 15th pitch with a 5.12a layback. It has gone at 5.8 A1—but very slowly, and this is a climb where safety lies in speed. In recent years, it has generally been climbed mostly free at up to 5.11a, saving most of the aid moves for the key 15th pitch.

The approach involves a bivouac on the extravagantly pinnacled arête of the Flammes de Pierre (the Flames of Rock, named for its appearance at sunset), then several rappels and a dash across a rockfall-strafed snow and ice couloir to the foot of the pillar. Descending this couloir is suicidal, so the only way out is up—nearly 2,000 feet of steep rock to the Shoulder, just below the summit of the Petit Dru.

Now, at last, the forecast called for a settled week of sunshine. I had just read the Bonatti Pillar episode in Joe Simpson's *This Game of Ghosts*, in which the author vividly recounts how the ledge on which he and his partner were bivouacking fell off the mountain, leaving them trapped all night, pupa-like in their sleeping bags, while dangling helplessly from a disintegrating piton placement. This was a less-than-ideal choice of reading material, and I was not surprised to feel a little uneasy as we set off across the Mer de Glace.

At the tiny Refuge de la Charpoua, perched high above the glacier, the surrounding granite spires were cloud-bound. The *guardienne* took our names and route, and wished us luck.

From the refuge, the approach follows an icy ridge before crossing the Glacier de la Charpoua, which was in an alarming state. Wet snow draped the rotten seracs. One snowbridge was like a wafer of Swiss cheese. Rotten snow, broken ledges and a gully brought us to the Flammes de Pierre. Two parties passed us on their descent from the Bonatti Pillar. They had set times of 12 and 14 hours, and looked more dead than alive. I wondered if going that fast was worth it.

Our pot bubbled on the stove as the sun set, turning ragged flags of lingering cloud neon pink. The pillar, like elephant hide in shades of red and purple against the velvet evening sky, towered toward the first stars and plunged equally impressively into the darkness, making my stomach tighten. Hugo was in rapture.

"I can't believe it. Worked my guts out for this, and now I'm actually up here, seeing this." He seemed close to tears.

What to leave behind? There was a lot of snow, so we each took plastic boots, crampons and an axe. We had four liters of water, cheese and some chocolate, and a minimal rack. I packed a light down parka and waterproof jacket, and wore thin gloves, trousers, longjohns, shirt, fleece and a balaclava. We also took my one-man bivi sack and Hugo's light down bag.

We halved the guidebook time for the approach to the couloir. The final gripping free rappel reached the snow only on rope-stretch. We scampered across the cratered névé, and down to the pillar's rotten base. Though the mountain was eerily silent, we were glad when the first easy pitch was behind us.

I'm still not sure where we got off-route. The third pitch—the Green Lizard Corner—was nothing like the 5.7 suggested in the guide, though it was festooned with rotting slings. The loose corners above deviated from the guidebook description just enough to be unsettling, yet they did not offer any alternative line. Pegs and slings were present, but fewer than expected. Things reached a head after the seventh pitch, when Hugo said, "This can't be it," and climbed back down. He tried up right instead, described it as his hardest-ever lead, and rappelled back.

"Two pegs with abseil slings, then completely blank," Hugo said. The pitch was meant to be 5.7. "We have to be off-line."

We rappelled down, looked right, looked left, but found nothing. We ate cheese and debated.

"I think all we can do is start again," I said.

Hugo looked appalled.

"What do you suggest then?"

We rappelled back to the first pitch. Nothing.

Sunset found me on the square top of a big flake—Hugo's previous high point—with Hugo struggling interminably through the blood-red overhangs above, aiding on our handful of nuts. He disappeared with the sun, reappeared an hour later, and advocated a bivouac on my flake. It wasn't the most comfortable night. My bivi sack stretched just past our waists, our legs dangled, and my back, hunched against the cold cliff, grew sore. The floodlit Aiguille du Midi hovered like a ghost above Chamonix's twinkling lights. We shivered a lot, slept little. By sunrise, action seemed very attractive.

Four parties had rappelled into the dusky approach couloir. Some Americans were battling a jammed rope. A French pair was already a ropelength below us, well to the left.

"That's the route!" I jabbered at Hugh, who was stoically repeating his evening's efforts overhead. A diagonal rappel brought us to a ledge below a chimney. Everything made more sense, and I felt better immediately, but then Hugo bottled out of leading. During the night, I had discovered, he wore only a light pile jacket with a shell and a thermal top. His lightweight trousers had ripped, baring his legs and bum. I hoped he would rally as he warmed up, as the hardest was yet to come.

I led the pitch, finishing in an unprotectable crack. "Thanks for taking that one," said Hugo, who then virtually ran up a vertical flake that wobbled alarmingly. I could see that "the Platform," three-fifths of the way up, was only a pitch away. From there, I told myself, routefinding would be easier. Crumbling holds led to a hideous offwidth, choked with rubble and decaying ropes, followed by a chimney that punished me for wearing a rucksack. I groveled gratefully onto the sunlit pedestal above.

We consumed most of our remaining rations, and were joined by three cheerful Japanese. Hugo fled up the next pitch, now hotly pursued. Menacing cauliflower clouds were boiling up around the massif. The forecast couldn't be that wrong, I thought. But it was.

"Hugo, I think we're in trouble."

I followed our topo with difficulty through the overhang above. One of the Japanese came up just behind, hesitated, then followed a crack rightward. Bugger the book, I thought, and joined him.

The rock was loose but pegged. Hugo joined us on the heels of the Japanese climbers, who introduced themselves as Go, Hiro and Ozaki. The sky was a sickly bruise color, and Mont Blanc wore an ominous cap of smooth cloud. Continually climbing over each other was holding us all up. Very worried now, I hesitantly asked the others if we could climb together to save time; the third person could bring up the last two while one of the first two led on. Unfortunately, their English was broken and my Japanese limited to "hai!" (yes). Eventually, however, they indicated we might join forces if the weather got worse. I suspected it was no longer a question of "if."

The next pitch was harder than expected, and Hugo and the others resorted to aid. Following, I could hear distant booms like monstrous footfalls. Wet spots appeared on the rocks and began to coalesce. Something dark and formless was swallowing the valley in huge gulps. The Japanese were rummaging in their sacks. Then it was on us.

All three Japanese lined up neatly along a tiny ledge in a lightweight bivi shelter. In contrast to this display of Japanese efficiency, Hugo and I wrestled in the deluge with my bivi sack, which vigorously resisted going over our heads. A flash made us cringe—thunder followed a split-second later. "Shit, that was close." Dangling from tethers, we couldn't cover our legs. Soon they were soaked.

It was clear we couldn't continue like this. Fingers numb, I rappelled to a ledge which promised some shelter. Hugo descended stiffly, drenched but uncomplaining. Then the deluge eased, and back we went, hopeful of climbing. It came and went, teasing. Eventually a Japanese head appeared, saying, "We climb together, yes?"

Go had racked up. Trailing an impressive collection of etriers and slings, he took a visible deep breath and began an A1 crack pitch. Hugo dangled motionless from the belay, eyes dull.

"Are you all right?"

I had to ask twice. A look of puzzlement crossed his face. "I think I'm abseiling."

Eh? I wasn't sure how to deal with such a bizarre answer.

"Well, here, tie in."

I put a half figure-of-eight in a rope end, and he fumbled with it as I tied onto the other rope. Something made me check his handiwork. He had not tied on.

"Always thorough, me," he slurred.

For a moment, I just gaped. I grabbed him by the shoulders. "Hugo, tell me exactly how you feel."

His eyes wandered. "Dunno."

"Hugo! It's very important. Think very carefully. Then describe exactly what you feel like."

I didn't know much about hypothermia, but his slow, muddled description confirmed my worst fears. He couldn't feel hands or feet or legs, had stopped shivering, and felt cozy now. He felt certain he was rappelling. His trousers were rags about his knees. He didn't even have a balaclava. I ripped off mine and, firing incessant questions at him to start his brain ticking over, plonked it on his head. His legs and forehead felt like a corpse's.

"Have you a poly bag?"

Blank stare.

"A waterproof bag." No. I fished out my trash bag (I had all my clothes on anyway), tore neck and sleeve holes in it and, taking his rucksack, ordered him to wear the bag. Before I could stop him he had put it on upside down, shredding it.

"Hugo!"

The rain was torrential again. I pulled my bivi sack over his head and rucksack, but it would reach only his thighs.

"Hugo, listen. You must warm yourself up. Move your legs."

A feeble rocking started.

"No, hard, dammit. If you don't warm up and keep yourself warmed up until we get off this thing, you will die! Understand?"

I had no idea if it was the right thing to say, but the rocking became more vigorous.

"Move your arms. Get angry. Punch. There must be something you can get angry about."

"Plenty."

I pulled up the bivi sack and rubbed his torso and thighs until I was breathless and pumped. "How d'you feel?"

"Dunno."

"No warmer?"

"Numb level's going back up my legs."

I rubbed them until I felt faint, then turned back to his body. To warm his body core, I really shouldn't have been rubbing his legs, but, if his legs were numb, he would never climb anything. At least my frantic rubbing was keeping *me* warm.

"Now?"

"Feel my knees now."

"And your core? Can you tell any difference?" A pause.

"Cold . . . but better. I'm clearer now."

I began to breathe again, but kept rubbing, coaxing and bullying as the rain poured onto us. What the hell was taking Go so long? Hugo still felt like ice. As I gave the pitiful figure some of our last chocolate, I found myself reflecting, with a pang of guilt, that he was only 19. It was all so unfair.

"Legs are back," he said eventually, jerking rhythmically on his leash. "No feet, though."

Seven more pitches to the Shoulder. The storm appeared to be fading. Perhaps I could just keep Hugo going until he dried off.

It was an eternity before the others were up the pitch and Hiro yelled from above that we could climb.

"At least there's no one to take incriminating photos," I said, almost producing a smile. My hands had warmed, causing excruciating hot-aches. Hugo steadied me as I moaned.

"You first," I said.

"Can't do it," he wailed. "Legs won't work."

"Get angry. You hate this pitch. Give it hell!"

He actually roared, thrashing upward until a free passage had him again protesting he couldn't do it.

"You have to!"

The climbing was strenuous, hugely exposed and partly overhanging, but Hugo was going as fast as me as we climbed together on the two Japanese ropes. I dropped a sling, fingers numb again. The climbing seemed endless, and I kept getting a sensation of déja vu, as if I were climbing the same pitches over and over again. At some point there was a distant boom. Then another. And another, each louder than the previous one.

The second storm hurled up the valley like an enraged ogre—a black, amorphous wall that exploded upon us in a chaos of wind, flying water, light and noise. We climbed for our lives, shivering at belays in the howling bombardment and premature night, skidding on half-seen, verglased rocks. I found the others in a walled niche under a small overhang. Our world was a battleground of dark and light. Rain, hail and sleet were flying everywhere with a ferocity that was astonishing. I figured we were just below the Shoulder, but we were too vulnerable to keep moving.

We wedged the metal gear outside and sat on our saturated ropes. The niche would have been tight for two, but five of us crouched there inside the Japanese bivouac sack, through which streamed deluges of spray. Hugo and I put on socks and plastic boot inners to try to recover our feet. The bivi sack thrashed as though alive. Why didn't it tear? My parka was saturated, as was Hugo's down bag. I was soon shivering. Worryingly, Hugo wasn't. He just sat there. Ozaki produced a stove and some soup, which raised our spirits.

Suddenly, though, a flash and detonation filled the niche. I felt as if the bones were exploding inside my legs. Screams of shock and pain mingled with the storm's cacophony. Then, Hiro and I were clutching ourselves, moaning, as the rest looked on with popping eyes.

"That," said Hugo, "went right up my arse."

Minutes later, we were all struck again. And then again. And again. After the first ten, it seemed almost tedious, until one particularly painful detonation left the niche full of screams of agony. When I regained consciousness, my knees felt like they were being ripped off, and my right lower leg was totally numb. Stuck in the entrance, I feared I was getting the worst of it. I wanted to ask someone else to take his turn, but that seemed peevish. A sickening stench of burning filled the bivouac. All eyes bulged with terror.

I was electrocuted 17 times that night. I hugged Hugo, trying to warm him. I wondered if the lightning was helping in that respect. Would things be better in the open? At least we wouldn't be stuck in here being fried. Out there, though, we risked being blown from the pillar.

It was with weary surprise that we greeted morning's light. The storm raged on. By mid-morning, though, the lightning abated, and we were ready to climb in a minute flat. Slush lay in pools on ledges, and cracks lay under a frothy skin of flowing water. The Japanese thought we were at the Shoulder, where the descent to the Flammes de Pierre begins, but for once I was sure my instincts were right and insisted we continue upward. Go led the last pitch. Hugo hadn't strength to follow and jumared up his rope. We both wore plastics on still-numb feet, and I scrabbled up the slippery rock, burning energy at a profligate rate.

At last, the descent. As if to spur us on, more lightning exploded only yards away. The Japanese insisted on bulky knots in the rappel ropes, and I was convinced they would jam. But I was glad to have their quietly competent company and to share the leading of nasty sideways rappels, which were murder in big boots. We now were working with feverish coordination.

Again and again I heaved the rope down from rappels, hands cramping, fingers like hideous, insensate white sausages. There was a flash. Electricity ripped up my arms from the rope. My skull seemed under pressure, and awareness splintered like a broken picture. My throat was raw. There had to be an end to this.

In the next hour, Hiro and I were struck twice more. I had no idea how this was affecting me, but my first job was to get down. My mouth felt like sandpaper. One more pinnacle, just one last big effort. The pinnacle was terribly exposed. We'd have to move together over it, rapidly. . . .

■ ■ ■

Hugo sat in the Charpoua Hut while I made him tea. I watched him shiver, violently and uncontrollably, hardly able to believe he had kept going so long. I had hoped to put him in the down bag we had cached on the Flammes de Pierre, but found it waterlogged and useless inside his bivi sack. Then came a jammed rappel rope. Then a snow bridge collapse, forcing a scary detour. And then, just a kilometer from the hut, a storm of huge hailstones drove us blindly toward the cornice above the glacier. It was almost funny.

The storm thundered on in the darkness outside. We had made it. Go, Ozaki and Hiro, who had descended ahead of us, were dazed but talkative. Friendships had been forged. The omelette cooked by the *guardienne* after she radioed mountain rescue was like an injection of life, although it gave Hugo stomach cramps.

As the evening wore on, and the groups behind us failed to appear, the mood became more somber, and I lay in my cozy bunk thinking of them. I wondered who they were, and if any had already lost their struggle. The prospect of another night out there was more than I could imagine. Later that night, my whole body was seized by agonizing cramps, and Hugo had to feed me salty water.

As I walked down in sunshine early next morning, helicopters plucked the seven other groups caught on the Dru. Soon after we had left Chamonix for the climb, a weather bulletin had warned climbers to stay as far from the mountains as possible. Miraculously, no one on the Dru was killed. In other parts of the massif, however, five lost their lives.

Hugo's nightmares began immediately. He dreamed of me being awakened by electrocution. Screaming.

Over and over again.

playing god on denali

Michael G. Loso
May 1998

How could a nuclear physicist be foolish enough to attempt one of the coldest mountains on earth in nylon hiking boots and the kind of rectangular sleeping bag that has little ducks printed on the red flannel lining? Mike Loso asked himself that question after witnessing Victor Pomerantsev's ill-considered attempt on Denali, which nearly ended in tragedy. This fascinating story was the result of Loso's thoughtful inquiry.

There wasn't much room for pacing, so Roger Robinson stomped three feet, threw his walkie-talkie down on a bench, turned, and stomped three feet back.

His mustache and eyebrows were frozen solid; melting snow dripped off his back and arms. In the heated confines of the little medical tent, he literally steamed. Spindrift poured in through the half-zippered door, and the occupants could feel the pressure in their ears and chests each time a wind gust leaned into the stout steel frame of the tent. Pots rattled on the stove, spices and teas tumbled off hanging shelves, and the plywood floor bounced with the pounding of the mountaineering ranger's heavy plastic boots.

"Dammit!" Robinson gasped, turning again. "Someone get on the radio and tell those guys that Victor is just going to have to wait. I'm not going to kill myself to rescue that fool!"

■ ■ ■

That "fool" was Victor Pomerantsev, a 54-year-old nuclear physicist from Sweetwater, Texas, with 30 years of climbing experience, including five 7,000-meter peaks. Despite this impressive background, Victor had attempted to climb Denali, one of the coldest mountains on earth, with nylon hiking boots, no ice axe and the kind of rectangular sleeping bag that has a metal zipper and little ducks printed on the red, flannel lining.

By the time Victor was flown off Denali in a rescue helicopter, he had jeopardized the safety of a dozen persons, scrapped the summit hopes of two climbers, and cost the federal government more than $10,000. In some years, taxpayers spend almost half a million dollars on search and rescue activities in Denali National Park. The public is increasingly scrutinizing such costs and asking whether the climbing community should be held liable for the foolishly bold in its ranks.

In such a climate, one has to wonder, couldn't expensive and dangerous rescues be minimized by simply refusing access to the unprepared? Perhaps, but who decides who is "prepared?" Was Victor?

■ ■ ■

Victor Pomerantsev was born in Khar'kov, Ukraine, six days after Russia entered the Second World War. When Germans invaded the city, his family fled to Siberia. He ultimately returned to Khar'kov, where he obtained a Ph.D. in nuclear physics and began working at his lifelong specialty: laboratory testing of radioactivity detectors under extreme environmental conditions—especially severe cold.

Victor started climbing, and he soon sought out the big, snowy peaks of the Pamir and Tien Shan mountains. For over 20 years, Victor spent most of his spare time in the mountains; eventually, he climbed all of the highest summits in the Soviet Union, including Peak Lenin in 1969. He was a candidate for the title of "Climbing Master" and once served as the Chief of Safety for an international mountaineering camp on Peak Victory. Then, in 1992, after the collapse of the Soviet Union, Victor found himself free to emigrate. At the age of 51, he moved to the United States.

There, in the oil and sagebrush country of Texas, he was suddenly paid handsomely to do the same work he had performed for five dollars a month in Khar'kov. He bought a house and a car, and, in his first act as a bona fide American climber, he went on a road trip. Victor drove to the Chisos Mountains in Big Bend National Park, and then to the Guadalupe Mountains, a small range in west Texas. It was pretty, he says, but the desert peaks felt as alien as American culture.

"My spirit was hungry for a white mountain," Victor recalls.

Denali was an obvious choice—it was high, snowy and, from what Victor had heard, easy if the weather was good. In April 1994, he contacted several guide services but was disappointed that none would schedule a West Buttress ascent for less than 20 days. With no American climbing partners, and only 10 days off from work, he decided to climb Denali alone.

Victor was not ignorant of the dangers of solo climbing—he had seen many good climbers killed in the mountains, and his first mountaineering instructor died during a solo ascent of Peak Victory. But Victor was convinced that there was no alternative to a solo climb, so he settled on what he thought was a conservative plan. He would try to "run up" Denali in four or five days, but would retreat at the first sign of a storm.

Climbers with experience in the high mountains of Europe or Asia often underestimate the ferocity of Denali's weather. They may not understand that when a storm hits, you can't simply "run down" to safety. Many are also unaware that air thins more quickly with altitude in the polar regions, so the physiological challenge of climbing 20,320-foot Denali is equal to that of a Himalayan peak several thousand feet higher.

On June 11, the day before Victor boarded a small plane for the flight from Talkeetna to the southern slopes of Denali, mountaineering ranger Daryl Miller had a hard time convincing Victor of any of this. Rangers ultimately have no authority to stop climbers like Victor, so Miller could only listen as Victor tried to explain, in broken English, that he was an experienced high-altitude mountaineer, fully capable of climbing Denali alone.

The next day, as Victor walked out of basecamp at the makeshift airstrip on the Southeast Fork of the Kahiltna Glacier, onlookers might have doubted his experience. He was wearing a light cotton sweatsuit, and his old army-surplus backpack was rotting. Its canvas shell was held to the rusting external frame by shoestring, tin buckles and frozen cotton straps. The pack held almost none of the equipment a modern mountaineer carries on Denali. For protection from the weather, he carried an uncoated nylon windbreaker, a spare sweatsuit, sheepskin-lined leather work gloves and a stocking cap.

Victor packed a pair of homemade crampons, a short rope and a hammer more appropriate for carpentry than for climbing. He brought no ice axe. Neither did he carry a shovel, thinking he could, if necessary, dig with the small bowl he brought for eating. He carried only four gas cartridges for his small butane stove. He planned to eat *spec,* a traditional Russian climbing staple that resembles, to the western climber's eye, raw bacon.

Victor brought neither skis nor snowshoes, an enormous gamble because summer storms routinely bury the packed trail on the lower Kahiltna Glacier under several feet of snow. To compound his dependence upon good weather, Victor wore cotton socks under a pair of lightweight Gore-Tex hiking boots. He fashioned a pair of homemade overboots from the same vinyl used to surface barstools. John Grieve, who eventually saved Victor from likely freezing to death, says these overboots looked "like something stitched together by a cave man."

"I understand John's surprise," Victor says. "My equipment was not good. That is true. But I believed in this equipment."

Like most of his Russian climbing peers, Victor had learned to make do in the mountains with much less than even the dirtiest of western dirtbags takes for granted.

Indeed, Ranger Miller acknowledges, "Some of the baddest-ass climbers come from Russia with the worst gear in the world." Miller recalls a Russian soloist that in 1992 climbed Denali in four days with a duffel bag for a backpack, smoking cigarettes and drinking coffee at every rest break. "He was the most tattered individual I've ever seen."

Victor's first three days on the Kahiltna Glacier were beautiful, and he found "very easy walking." But instead of traveling at night, as many climbers do, to cross deadly crevasses while the snowbridges that span them are frozen solid, Victor traveled by day. He sweltered in the intense June sunlight—growing more dehydrated each day—and he found it very cold at night. He began crawling into his thin sleeping bag with his boots and clothes on.

"I began to feel a very big troubling," he says of his experience. The cold temperatures scared him, and the easy traveling, too. "When things start easy in the mountain, they always get bad," Victor claims.

Victor was not fearless. He had no death wish, and the magnitude of his undertaking genuinely scared him. But lying awake at night, staring into the cold twilight, he reassured himself that he would simply retreat if conditions got dangerous. The walking, after all, was very easy.

■ ■ ■

Victor reached his highest camp on June 15, four days into his climb. Most Denali climbers stop for several nights at Genet Basin, at 14,200 feet, which provides a convenient and relatively sheltered advanced basecamp for summit bids. But Victor passed this camp during the day and climbed alone up the steep face that accesses the upper part of the West Buttress.

If Victor had stopped in Genet Basin, he might have heard about the big storm that was predicted for the coming week. He might have spoken with the search and rescue team stationed there and been convinced that his gear was inadequate for the conditions on the upper mountain. But, according to Victor's plan, he must try for the summit on his fifth day. So, while at least 50 climbers in Genet Basin prepared for the storm by strengthening their campsites with snow blocks and walls and caves, Victor set up his pup tent on an exposed ridge at 16,200 feet, with fantastic views of the Kahiltna and Peters glaciers thousands of feet below.

During the night, a small storm blew in. Twenty to 30 mph winds buffeted and buried Victor's tent in almost two feet of spindrift. In the morning, however, the clouds lifted and it appeared the weather might improve. Victor decided to rest for the day and hope for a summit opportunity the next morning. Down in Genet Basin, camp buzzed with the possibility that this might be all that would materialize of "the big storm." At least one climbing party took the bait. Bill Ross and John Grieve, two climbers from Colorado, grabbed a week of fuel and food and set out for the 16,200-foot camp, hoping for a summit opportunity.

Bill, 40, is a backcountry ranger in Rocky Mountain National Park, and John, 39, is a district forester for the state of Colorado. Neither man is a strong technical climber, but they approached Denali with extensive experience in winter backcountry travel. For a year before their trip, they practiced crevasse rescue, read books on altitude illness, mountaineering, and Alaskan survival skills, and acquired what they felt was the best gear possible.

They climbed Denali cautiously and methodically. By the time they reached the camp at 16,200 feet, they were on the 12th day of their expedition, well-acclimatized and ready for a shot at the summit. At the top of the fixed lines, however, the winds were again starting to blow.

Bill recalls seeing Victor's tent, still half-buried from the previous night's storm. But it was so limp and ragged he thought it was just an old cache of clothing, or perhaps food dug up by ravens. The wind was already stretching clouds around the summit into hazy spider webs, so he and John hurried past Victor to set up their own tent.

That night, the real storm arrived. As the Coloradans took turns shoveling out their tent, they threw the snow into a steady 50 mph wind. It carried the snow south, across Victor's buried tent, to pile up in a huge pillow on the headwall above Genet Basin. With five days of food and fuel, a strong tent, and snow walls three blocks deep, the two felt confident and safe.

Only 30 meters away, Victor was being buried alive. He sleeps well in windstorms, and this fact worked against him. Late in the night, cold pressure against his face finally stirred him. He opened his eyes and saw that the roof and walls of his tent were completely collapsed. He tried to push outward on the tent walls to clear them of snow, but they were packed solid.

"It was a very awful picture," he recalls. "This night, it was dark in the tent, because of the very big snow, and I didn't realize until I got out how much snow was on top of my tent."

When Victor crawled into the brunt of the storm, he saw that his backpack also was buried, and it contained his gloves and his bowl—his only digging tool. "I tried a few times to dig them out, but my hands would not work in these conditions," he says.

Had Victor known that Bill and John were only meters away, he could have gone to them for help. But he didn't see their camp through the storm, so he crawled back into the tent. He remembers lying there for 10 or 20 minutes, trying to warm his hands and watching the tent collapse even farther. He says this was a very "mystical" time, adding, "I felt the presence of death, which I had never felt before. I felt her breath on me."

Then, as Victor tells it, he reached a turning point. Remembering his family and friends, he decided to fight for life. He grabbed his cotton sleeping bag and rolled out as the tent finally collapsed. Trying to shelter himself in the lee of the drifted snow, he wrapped himself in his bag and turned away from the howling wind, trembling and straining to stay warm. Gradually, though, he relaxed, and Victor describes the next eight hours as a very calm time during which his training and discipline allowed him to triumph over the elements.

"I was sure some climbers would come my way in the morning, and that I could ask for their help digging out my equipment. Maybe, I thought, I could go up if the weather was better the next day."

Bill Ross, who found Victor that morning, recalls the story differently. "At about 8 a.m., I got up to dig out the tent and saw something sticking out of the snow. It was his cheek and one ear, sticking out of the snow, frosted. He was half out of the tent and half out of the sleeping bag, kind of curled up in the snow—kind of fetal. He had no gloves or hat, and his fists were clenched by his cheeks. I brushed him off and got no reaction. I thought he was dead, but when I pulled on him, he was semi-flexible."

Bill called to John, and, while the storm continued, the two carried Victor to their tent, stripped him of his wet cotton clothing, and shoved him into the sleeping bag with John while Bill started melting snow to make hot water. Several minutes later, Victor was still too cold to shiver, and his oral temperature was only 88 degrees. What Victor remembers as a lucid, calm time was really hypothermia and a complete loss of consciousness.

"If Bill hadn't been out there shoveling," says Roger Robinson, "Victor would be just another cross at the cemetery."

■ ■ ■

Bill and John radioed for assistance later in the day. By that time, they had rewarmed Victor in their tent, and he was no longer in any mortal danger. They just wanted to be rid of the Ukrainian so they could continue with their expedition.

Roger Robinson and three members of the volunteer rescue team set out through a wand-to-wand whiteout to try to reach the 16,200-foot camp. The four rescuers roped up and waded through waist-deep snow and 90 mph gusts to the base of a steep headwall that led to Victor. High winds on this exposed headwall typically scour the névé and blue ice to reveal a pair of fixed lines. But, pawing around blindly in the snow, Roger wasn't able to find those yellow polypropylene lines. It was only then that he sickly realized what was happening. This storm had blown in from the north, scouring the other side of the ridge and loading the headwall above him with over 1,000 vertical feet of steep, unstable windslab. He had just led his volunteer rescue team into the path of a very big, loaded gun.

The crew made it safely back to camp, but Roger was pissed. Despite 14 seasons on Denali and a firm commitment to the first rule of rescue—don't become another victim—Roger had been tricked by the mountain, and he knew he was lucky to walk away. The rescue had to wait. The news was delivered to the Coloradans by CB radio: Victor was theirs for the night.

Static didn't hide Bill's frustration. The Ukrainian was uncooperative, he said, and ungrateful. Helpless and completely dependent, having lost virtually all of his gear and frozen his fingers, Victor nonetheless claimed he was in no need of rescue. Resting cheerfully in Bill's sleeping bag (in which he had defecated while recovering from severe hypothermia), Victor said he still hoped to climb the remainder of the mountain if the weather let up.

At this, Roger exploded. "That son of a bitch is sitting in their tent, eating their food, shitting in their sleeping bags, and claiming he still wants to climb the mountain? We could have been killed just now, to save this fool's fingers! Who does he think he is?"

■ ■ ■

To hear Victor tell it, Bill and John forced him into their tent, when all he wanted was a little help shoveling out his gear. On the fourth day of the storm, when Victor's hands had swollen to twice their normal size and were covered with giant blue and black blisters, he demanded several times that the Coloradans walk him down to the medical camp. Or, maybe they should just let him go down alone. By that point, they badly wanted to be rid of Victor. But the storm outside was deadly, so they prevented the Ukrainian from walking out of the tent to his death.

On the evening of June 21, six days after the onset of the storm, the clouds finally cleared. A helicopter evacuation was attempted the next morning, but not because Victor's life was in jeopardy. Says Robinson: "We were flying him off to protect others. He was a liability to everyone else on the mountain."

When the helicopter was unable to land on the ridge because of continuing gusty winds, the frustrated Coloradans began breaking down their weather-beaten camp. Later that afternoon, they tied Victor to the fixed lines with a short rope, wrapped his hands in gauze and giant mitts, and began walking him unsteadily down the headwall.

At the base of the fixed lines, they found an unexpected obstacle. There, at the point where deep snow had turned back Roger and his rescue party four days before, the three men carefully downclimbed an eight-foot-high fracture line left by a massive avalanche. Sometime during the storm, a 500-foot-wide slab had swept to the base of the headwall, less than 200 feet from the tents in Genet Basin. The sight of it made Roger wince.

Bill and John arrived in Genet Basin as minor celebrities, as everyone trapped there during the storm had followed their story closely by CB radio. But there was little satisfaction in the congratulations from climbers who were busily packing up for a series of fine summit days. With their food and fuel nearly gone, Bill and John had to leave. And Victor's rescue had cost them more than the summit. Over the past few days, Bill's fingers had grown numb and had begun to tingle, and he knew that he had frostnipped his fingers while melting snow on the first morning they found Victor. Camped that evening near the medical tent, John wrote in his journal: "Defeat."

■ ■ ■

Victor was flown to Anchorage the next day. Four months later, a surgeon in Texas removed the first digit of seven fingers and most of an eighth from Victor's left and right hands. Air conditioning makes his hands hurt, and his heightened sensitivity to cold prevents him from spending time in the mountains. "Life continues," he says.

For Victor, of course, life continues only because Bill Ross and John Grieve saved it. Today, Victor understands this: "I feel very big grateful feeling. I was dependent on them in everything. I understand now that they keep my life."

But, for Bill, there is little satisfaction in heroism.

"When somebody is in trouble, you help. It's what differentiates mountaineering from life in the city—you don't just walk by," he says. "But Victor was an accident waiting to happen—that's the aggravating part. I've never seen anyone as blatantly underprepared. I don't necessarily think it's the Park Service's job to screen climbers, but Victor should never have been in that predicament."

With hindsight, most people would agree—Victor had no business on Denali in 1994. But should the Park Service have stopped Victor before he even got on the mountain? Like Bill, most people involved with the incident say no.

"Everyone should have a right to face these challenges and risks," says Robinson.

Daryl Miller, who registered Victor before the climb, says, "None of the flags went up with Victor, no 'ding-dong, oh-boy, dumb Russian climber' or anything. He had a strong climbing resume, and he could talk the talk."

Victor's important mistakes—planning too little time for the ascent and climbing with inadequate equipment—stemmed primarily from ignorance, and, to a lesser degree, from something very difficult to verify in a half-hour meeting with rangers: a simple lack of common sense.

Rather than expecting rangers to judge the preparedness of climbers, the best approach to minimizing rescues on Denali, and elsewhere, may be a more vigorous approach to climber education. This is particularly true for non-native English speakers like Victor. English-speaking climbers have access to books, pamphlets, briefings, daily weather reports and the "war stories" that circulate within the climbing community, both on and off the mountain. Denali National Park has been working to address this problem and recently published mountaineering-information pamphlets in eight languages, including Russian, as well as foreign-language videos.

However, language barriers alone are not the cause of accidents. In fact, the number of foreign climbers requiring rescue on Denali in recent years has been proportional to their total numbers. Between 1994 and 1997, about 41% of the 4,755 climbers attempting Denali came from foreign countries, and 41% of the 80 climbers requiring rescue (including evacuation of dead climbers) were foreign.

Denali National Park also has implemented a controversial requirement that Denali and Mount Foraker climbers preregister at least 60 days before their climb. For those who go to the mountains to get away from rules and bureaucracy, this seems an infringement on climbers' rights. But it does allow time for registrants of all backgrounds to thoroughly consider available information on these dangerous peaks. With a policy of education, however burdensome, the government can at least leave decisions about climber preparedness in the hands of climbers themselves, where it belongs. And the education of climbers may be working. Since 1994, the number of rescues has declined each year.

Robinson, who has risked his life over and over for the unprepared, thinks that giving the decision on who can climb to a government agency would be a grievous dilution of the wilderness experience.

"I'm not sure rangers should have the authority to say no," says Robinson. "The 'Victors' are not the cause of all problems—experience level has no correlation with accidents on Denali."

Chief Ranger J.D. Swed agrees.

"Denali kills just as many very experienced climbers as underprepared climbers," he says. "There's a lot of ways to be prepared or be unprepared on the mountain—some people have good skills and no money for gear, but probably a larger group has great gear and too little experience. We're just not willing to play God."

Better, in other words, to have a world where a few perish in the mountains, than one where only a few are allowed in the mountains at all.

new york stories:
bouldering in the big apple and beyond

Josh Lowell
April 1999

Bouldering can be great almost anywhere. You don't need towering cliffs—only a tiny, cast-off rock or two. And few boulders are lonelier than the ones of Harlem, in northern Central Park. The wild enthusiasm that Josh Lowell and his pals bring to these desolate rocks (and to the much finer stone of the nearby Shawangunks) casts a unique spell.

New York has taught me everything I know about bouldering. Here's what I know.

I. Every boulder problem begins in the mind. *The most compelling boulder in the world is merely a chunk of rock until someone notices it, imagines a path of human motion over its surface, and makes the critical leap of actually trying it. Putting on shoes, taping down fingers, chalking up hands and thrashing the body in the blind hope that something inspiring will reveal itself in the process. If you are a boulderer, if difficult movement on rock is your passion, you will find inspiration each time you open your eyes.*

I admit that I was doubtful when Ivan began raving to me about the incredible bouldering in the Gunks. I was living back on the East Coast for the first time in five years, spending the winter working in a gym between fall and spring road trips. At the competitions that year, the talk in iso was always the same, and it was always coming from the prolific mouth of Ivan Greene.

"Dude, the shit is rid*iiiiiiiii*culous, yo! There's *soooo* many fat problems, and *maaaad* hard, too. Full-on potential, *un*tapped. I'm *telling* you bro, it's the bomb!"

Ivan has the energy of three people stuffed into a frame too small to contain it. His chest and arms bulge obscenely, as if his intensity might burst through and explode across the room. When describing his boulder problems, his excitement becomes uncontrollable. His eyes shine, his fists clench, he grits his teeth and growls, pantomiming the cruxes. He searches for words to match his enthusiasm and ends up inventing new ones mid-sentence.

" . . . you slap to this gleck sloper, *whaaaaaatisch!*, then get into this total fungo position, like this? Then *Buhdoooogis!*, up to the most mackadocious pinch in the world. Then *dooooogis!* to this horrendous crimp, and *doogis, doogis, doogis* to the top. It's so *doooooooope!* You're gonna love it!"

A true New Yorker, Ivan's not given to understatement. In New York you have to shout to be heard, dress to be seen, and rush to arrive or you find yourself ignored, invisible, lost. My own tendency is to meet hyperbole with cautious skepticism. "Hmmm," I said to Ivan, casually nodding my head. "That sounds pretty good . . . " But as he continued his impassioned descriptions, I noticed a tingle of excitement spreading through me. Various shoulder and forearm muscles were contracting involuntarily, and I accidentally chalked up while standing there listening.

A few weeks later, a route-setting delay at a competition in New Paltz provided a spur-of-the-moment opportunity for Ivan to show off his creations to the East Coast posse. The day was grey and cold, but we optimistically piled into his Saab and jetted up to the Trapps, stereo bumping Notorious B.I.G. beats so deep that we were bouncing on the seats, occasionally banging our heads on the ceiling. Though we were due back at the comp for finals in two hours, we were pretty sure they wouldn't start without us. All of the finalists were in the car.

Ivan screeched around the corners, pumping his fists in the air and shouting at the top of his lungs over the music. His words alternated between Biggie's rap lyrics and tidbits of beta for his most recent masterpiece, The Illustrious Buddha.

We tumbled out the doors like shell-shocked clowns from a sensory-overload circus car, and followed Ivan on a mad-dash tour down the Carriage Road, stopping at classic problems established as far back as the 60s by legends like John Gill, Lynn Hill, Patrick Edlinger and Jerry Moffatt. Ivan tried to keep us moving, pausing just long enough to point out each problem, watch everybody fall a couple of times, then casually hike it with perfectly wired beta—he couldn't have been happier. Then he'd quickly throw his sneakers back on and hop around yelling, "C'mon, you gumbs! Stop flailing and send this thing so we can get to the *biz!*"

As we moved in fits and starts toward The Illustrious Buddha, we gathered quite a crew, and a sense of competition hung in the air like chalk dust in a gym. Even as one climber splattered on the crash pad, the next contender was eagerly hopping over him to grab the start holds.

By the time we reached our destination, a drizzle had begun and dark was setting in. Our skin was wearing down, we were shivering and the finals were supposed to begin any minute now, back in town. My first thought on seeing the boulder was, "Oh look, here's a nice big roof we can hide from the rain under. I wonder where the Buddha is?" My next thought was, "Why is Ivan putting chalk on that flat spot? Why is he pretending to lift himself up there . . . ? Wait a minute . . . "

Ivan stood under the huge, prow-shaped, horizontal roof. His arms were spread wide above his head, bear hugging the prow on opposing flat spots. He squeezed his arms together, straining so hard that his pectorals bulged to their full C-cup potential. He arched his back and piked his feet off the ground and into two small toehook notches. His body clung horizontally, with one flat spot and one toehook on either side of the prow. As all his friends and rivals stood around, too wide-eyed to spot, Ivan performed the bizzare and complicated sequence he had choreographed to climb The Illustrious Buddha.

I was blown away. I had looked at this boulder and seen only shelter—the thought of climbing it had not even occurred to me. This is not a boulder that begs to be climbed. It's not particularly high, nor does it have anything that looks much like a hold—just a series of acute and obtuse angles. But by imaginatively fitting the parts of his body on and around these angles, Ivan had created a subtle, dramatic and beautiful boulder problem.

When he finished, the rush began. Oblivious to the drizzle, the cold, the impending dusk and any thoughts of conserving strength for the finals, we threw ourselves at these fascinating moves while Ivan fed us the kind of genuinely encouraging beta that can only be supplied by someone who has just resoundingly sent the problem in everyone's face.

After numerous inglorious backflops, I finally managed to hoist myself up onto the four non-holds that start the problem. One move was about all I managed that session, but what an incredible move! My body had never done that before, and my mind could still hardly believe that it worked.

By the time we tore ourselves away, my brain was spinning fast, repeating that opening movement again and again, then leaping forward to future visits, dreaming up as-yet unimagined moves on unseen boulders. Ivan promised enormous untouched potential in the area, and my skepticism was waning.

II. From the minds of sick individuals come sick boulder problems. *To them, every new and difficult movement is a great one, and the hardest way is almost always the best. Gather a collection of such twisted people, and their intensity is magnified by one another through lenses of com-*

petition and camaraderie, then focused to a pinpoint on a single objective: one boulder, one move, one hold. The result is explosive.

Ivan, along with Marc Russo and a handful of other New Paltz and New Jersey climbers, had been establishing problems for about a year before our eye-opening session on The Illustrious Buddha. With that visit, a new group energy was ignited, and the pace of development exploded.

The converts included climbers from all over the Northeast. After gathering in The Bakery for coffee-chugging and rabble-rousing, we'd head out en masse, usually to the cliffs and talus fields of Peter's Kill, a section of Minnewaska State Park recently opened to climbing thanks to the efforts of the Access Fund. The boulders there, virtually untouched in the spring, held well over a hundred problems by fall.

On my first visit to Peter's Kill, I put up several obvious new problems. The movements were difficult and unique, and their realization sweet... for about ten minutes. Then on to the next! The thrill of newness kept me coming back all year, making the pilgrimage from Connecticut at least twice a week.

My mileage was outdone, however, by the driving exploits of Obe Carrion, whose six-hour round-trip journeys from Pennsylvania were fueled by a maniacal drive to send absolutely everything. The beginning of the season was back-to-school time for Obe. His obvious strength and talent were raw, fresh out of the gym, and he bashed himself against every hard move in the area trying to keep up with Ivan. By late spring he had learned the subtleties of Gunks rock, and the rest of the season became an ever-escalating frenzy of hard new development as Obe and Ivan strove to outdo each other, move by move.

Between problems, the two of them would race around the forest bugging out. Singing rap lyrics, attacking each other with *Mortal Kombat* moves—the most popular move, the Snake Style, was a two-fingered jab to the throat—and spotting intriguing boulders through the trees. They'd push each other out of the way to get the first crack at a newly spied line, then battle furiously for the first ascent. The beta was refined with each burn, a higher hold was reached, and the competitive fever rose.

"C'mon, Hector," Ivan would say. "You better send this thing right now, cuz I'm about to walk it."

"Who you callin' Hector, yo?" growled Obe, chalking his hands and mustering his psych. "Y'ain't walkin' nothin. I'mma show you how we do it Puerto Rican style." He'd put his head down and mutter to himself, "C'mon Obe, c'mon. C'mon Obe, c'mon!" then step to the rock.

Obe climbs to a self-produced soundtrack—a human beatbox of rhythmic exhalations that coordinate his timing and remind him to keep breathing. "F-T-PAAAA!" he shouts after sticking a deadpoint, "F-T-PAAAA!" While holding a deep, static lockoff and delicately reaching to a far-off hold, his breath escapes in a loud, slow hiss: "SSSSS-SSSSSSSSSSSSS . . . " And on reaching the hold, an explosion of relief: " . . . SSSSS-PAAAAAAA!"

He holds nothing back, and you can see the struggle in the contorted grimace of his face. He fully expects his body to be capable of anything his mind asks it to do, and those are often tall orders. "I feel I should be able to onsight every boulder problem I try," he's told me. "But sometimes my brain can see me doin' it before I have the physical capabilities of doin' it, and that's when I get upset."

Obe puts so much pressure on himself that outside observers sometimes wonder where the fun is. "Why don't you try to relax?" I've heard people ask him.

"Relax?" he replied. "Man, I can't be relaxin'. This is what I do. This is everything to me. If I was relaxin' I'd have no boulder problems done. I'd be lyin' on a beach somewhere, and I'd be relaxed."

With Ivan screaming "C'moooon! I got you! Go! I got you!" Obe would dig as deep as he could. When successful, his joy was unrestrained. He'd jump around on top of the boulder shouting nonsense victory cries in Spanish and laugh contagiously the rest of the day. Other times he'd run out of steam and crumple to the ground with a moan of exasperation. "Dude, you were there!" Ivan would cry triumphantly, hopping over Obe's fallen body for his chance to snag the first ascent. "Why'd you fall, you gumb?"

"I don't know, yo. I don't know. Just punch me in the back of my head," said Obe. "Then send that freakin' thing," he'd add. "It's easy anyway."

III. Group synergy can transform the humblest of boulders into climbing monuments. *After a project's been sent, the rock seems bigger, somehow more important. The chalked holds, like footprints across a deserted beach, provide a sense of human scale. The pain and frustration involved in mastering the movement is forgotten, leaving a completed and enduring thing of beauty: a boulder problem.*

It's two years later and things have changed in the Gunks—it's an area now. I recently ran into two strong, well-travelled boulderers out at Peter's Kill. They had driven from Colorado, planning to tour the eastern half of the US for a couple of months. But after bouldering at the Gunks, their first stop, they scrapped their plans to go anywhere else.

I could hardly believe it. I'd had a particularly mortifying climbing day, and I felt fed up with the tip-shredding, tendon-tweaking, sharp, little, hard-to-hold, goddamn crimpers. I'd had enough of the finicky conditions, torturous sit-down starts and head-and-coccyx-smashing talus landings. I just wanted to get the hell out of there and go to a "real" area.

"I think this place is just as good as Hueco or Fontainebleau," one of the guys told me.

When I found Ivan at another boulder and related the encounter, he was equally confused. "Are they smoking crack?" he asked. "Are they bugging?" Even Ivan, after three years of screaming the glories of Gunks bouldering to the world, had never dared such a sacrilegious comparison.

We dismissed the matter laughingly and went on bouldering, but I kept noticing Ivan slipping into a mood unusual for him: calm reflection. As for myself, I stared at the all-too-familiar boulders, trying to see them through fresh, outside eyes. Despite my frustrating day, I still had a deep respect for the bouldering here. But I also suspected that much of the reason I liked the area was because it was "ours."

"They're really not going anywhere else?" Ivan asked me that evening as we stood around the parking lot slowly stuffing our crashpads into the car. "I can't believe those guys."

"Yeah," I replied, "crazy." I was looking down at my feet, which were scuffling around idly in the parking lot gravel. A long silence passed, and it occurred to me that long silences almost never pass in the company of Ivan. After some time I looked up and added, "But it still makes me kinda happy."

Ivan broke out into a triumphant little foot-stomping dance, threw his head straight back and screamed up into the sky "WOOOO-OOOORRRRRD!!!!!!!!"

IV. Creativity and enthusiasm ensure great bouldering—the rock is a secondary resource. *People with minds demented enough to spend days torturing their bodies in the pursuit of horrendously difficult and utterly meaningless goals will find opportunities to do so wherever they may be.*

Besides being my main climbing partner, Ivan is also my boss at Chelsea Piers, a deluxe Manhattan sports club with a huge climbing wall. He hired me during the frenzy of bouldering development in the Gunks and coordinated our schedules so he'd have a partner every week.

At work, when we weren't training endurance or keeping the members from falling to their deaths, Ivan and I would relive the past week's bouldering exploits with the rapt attention of our fellow employees.

Preston Lear was always particularly attentive. Preston had moved to New York City from Salt Lake, resolved to the unavoidable fact that he'd be sacrificing his climbing habit for the rigors of the big-city academic life. He'd been surprised to hear of the bouldering scene up in the Gunks, just an hour and a half from Manhattan, but with a busy school schedule, a wife and dog, and no car, he only managed to taste it for himself a couple of times.

Ivan and I would act out all the beta for the newest crop of problems, argue the names and grades, and compare sore muscles. Preston, who hadn't touched real rock for months, would spend hours soaking up vicarious bouldering thrills. He was desperately in need of a fix, and besides, anything was better than loading the dishwasher with dirty holds or spraying down the rental shoes with Lysol.

During one of these bullshitting sessions, Preston arrived at work practically skipping with excitement. "Dudes," he said, "The Gunks sound great and all, but today I've seen my future, and it's a boulder in Central Park."

"You mean Rat Rock?" yawned Ivan. "That shit is played."

"No no no!" Preston exclaimed. "This boulder is a little bit off the beaten path. It's way uptown, actually. Actually," he stammered, "it's in Harlem."

A small bunch of New York City climbers had known about the spot for a while and had done a few moderate problems. But Preston had become obsessed with the idea of a traverse. Over the following months, we heard about the bewildered looks he got during his uptown subway rides with a crashpad on his back, though he assured us that, while the neighborhood was run down and dominated by housing projects, he felt totally safe bouldering there. Particularly when accompanied by Roman, his 120-pound Rottweiler.

Preston was making gradual progress on the traverse, and it was shaping up to be hard. We noticed him training more. Seems he got in the habit of making dawn pilgrimages to his beloved boulder to squeeze in a couple of burns on the project before his 9 a.m. class. "You guys should come check it out," he'd say, but somehow the invitation seemed half-hearted.

That changed as soon as he sent the traverse. No longer protective of his cherished project, he went into full hype mode, bombarding us with beta and cooing, "It's soooooo goooooood!" Ivan and I were reluctant to make the journey all the way to Harlem for one little traverse. We were both on rest days preparing for "important" projects in the Gunks, and besides, how hard could this thing really be? Did he really want to drag us all the way up there just to watch us rudely flash his pride and joy?

"Dudes," he said, with a knowing twinkle in his eye, "I'd love to see someone flash it." He shook his head, preparing to deliver the clincher, "But I really don't think it's possible."

So off we went, with Ivan driving and Preston waxing poetic about the Harlem boulder. "It's actually really high quality. But more than the climbing," continued Preston, who's studying for his masters in social work at NYU, "this whole thing's been a pretty intense cultural experience."

I'd long been picturing the super-solid rock and nicely textured slopers Preston was always going on about, but now he started filling in the rest of the picture. "One thing you guys should know," he warned us. "In this part of the park there aren't any public bathrooms . . . and, uh . . . I guess this rock provides the most privacy around."

"Oh no," I moaned.

"It's not exactly a pristine wilderness," Preston went on. "In fact, when I first started coming here the whole area was covered with broken glass, and there were crack viles, syringes and condoms all over the place."

"Fantastic! Why are you taking us here?" I said.

"What the dilly, yo?" said Ivan.

"It's not that bad now, though," continued Preston. "I met this ambulance driver here one day, and he gave me a box of quadruple-thick, needle-proof rubber gloves so I could pick up all the trash. I filled a whole garbage bag!"

"Damn," said Ivan.

"Jesus Christ," I said.

"Yeah, there's usually some sketchy characters around, too," admitted Preston. "But there's always cops nearby. Last time this cop car pulled up right next to the boulder and the guy leaned out the window and yelled 'Hey you!' I thought he was gonna hassle me about climbing, but—listen to this—he goes 'Watch yourself around here. We pulled a corpse out of the bushes behind that rock yesterday!'"

By now we were pulling up to a parking spot next to the edge of the park, across from a burnt-out building. It was too late to turn around.

"Preston," I said, "this better be good."

Preston zoomed ahead on his skateboard while Ivan and I lagged behind, scoping out a bunch of outcrops along the way. When we arrived at the boulder, Preston made us do every warm-up problem before he showed us his creation, Family Values.

As soon as I started fondling the traverse's holds, I was fascinated. There were all sorts of complicated slopers, the kind you can stand and study for minutes at a time. Tunnel vision eliminated all perception beyond the surface of the boulder, and I was transported straight out of Harlem. I had come up here expecting a silly, half-assed session on some obscure piece of rock, but as I prepared to try the traverse I felt a calm, intense focus sweeping over me. This was the real deal. I flashed the problem with perfect beta from Preston, and it felt as significant as anything I'd ever done.

After that, we upped the ante with a desperate four-move addition to the beginning of the traverse. Ivan was also in serious bouldering mode, and after devising a crazy heel-hook sequence, he was ready to send.

We gave him a good spot and fed him continuous encouragement. But as Ivan began the problem, a gang of punks stopped on the path behind us for some after-school fun. "Yo, white boys!" they screamed at us in prepubescent voices. "What the fuck you doin'?"

There were two of us, shirts off, kneeling on a little fuzzy mattress and reaching out toward the back of a third shirtless guy who was hanging onto that rock where people go to shoot up and take shits.

"You faggots or something?"

Ivan, who was in the middle of the crux, tried to maintain his composure. All he had to do was hit that next good edge, and he'd be in for sure.

"Yo! Faggot! I'mma put a cap in yo' ass!"

Ivan slowly unwound onto the sloper he had just crossed to, coiling like a spring beneath the good edge and looking right at it. He froze for a second, and shook with body tension, unbreathing and turning red.

"Do it!" we shouted. "C'mooooon!"

"Homo!" screamed the kids.

And Ivan sprung, his body exploding off the wall, his hand shooting up to snag the edge, his feet flying out in a huge arc. A tremendous cry of "Flacco!" drowned out all other noises.

"Flacco?" I asked him later. "Why'd you say Flacco? That means 'skinny' in Spanish."

"I dunno," he said. "I was just focused on doing the move. That's what came out."

Watching Ivan's performance on the first ascent of Sweat of the Rapist, Preston once again saw a vision of his future. Preston is a patient, determined climber, who picks an objective and stays with it to the end. When he had done Sweat, he moved onto another project, and several months later added Privileged, a four-move line up the center bulge on ridiculous nothings. By now, he guesses he's made more trips to this obscure little boulder in the middle of Hell than to any other climbing area. He's hauled out bags of hazardous waste, and hauled in bags of woodchips. And he's dragged dozens of climbers there, who, like myself, arrived horrified and left inspired.

Every boulder problem begins in the mind. . . .

climbers

penned in leavenworth

Alison Osius
May 1988

Writers often speak of a "sense of place," when they really mean a sense of the people in a place. For Alison Osius, who spent some time in the fake Tyrolean village of Leavenworth early in her distinguished rock-climbing career, the gregarious, off-the-wall climbers of eastern Washington enlivened her stay far more than the climbing.

Alan Watts did it with Smith Rock. He not only talked the place up, he asked passers-through to.

Plenty of people have done sales jobs. This is how it goes. You have a good route, a wanna-be classic. Maybe you made the first ascent. But if you want people to repeat it—to enjoy it, to help rate it, to keep lichen from re-claiming it—you have to advertise. Or let's say you're in a good but quiet area, and you'd like to have someone to climb with, perhaps some folks from other countries. You have to broadcast.

■ ■ ■

Leavenworth is the best little sales job in Washington. It starts with Katie Kemble, soul of hospitality, who has a wildly enthusiastic way of urging you to come visit. This technique works. I met Katie when we were both rappelling down the Central Pillar of Frenzy in Yosemite, and by the bottom she'd talked me into a visit. By the end of a subsequent phone call, she had me taking a summer job at Jim Donini's school, Leavenworth Alpine Guides, and staying in her little A-frame in Icicle Canyon, right in the heart of the Cascades. She promised, "You can see Snow Creek Wall out of the living room window and go bouldering by the mailbox."

I rode into Leavenworth on a bus. It was dusk, and at first I thought I was imagining things, because it seemed as if giant flowers were painted across the walls of the buildings. But, in fact, they were. In the early '60s, Leavenworth was a depressed town of vacant store fronts. It had grown up around a railroad and sawmill, big employers that had long since gone. But the feisty residents approached the community development division at the University of Washington. Its advice: make the place attractive to tourists; go for either a Western or a Swiss Alps theme.

Thus the locals, drawing mightily on a few roots a few of them have, created the Bavarian Village. They gave its architecture an Old World Alpine makeover, adding ornate flower-covered balconies, insistent bric-a-brac, and pillared street lamps; the women pulled on *dirndls* and puffy sleeves, the men *lederhosen* and felt hats with feathers. They baked pastries, created restaurants and stores. The climbing shop in Leavenworth is called Der Sportsman.

The Bavarian idea worked. Today tourists stream through the town.

Just outside town is Icicle Canyon, one of North America's deepest gorges with its 6,000-foot vertical relief. Far above town, cliffsides lace the mountains; closer, they dot the hills and line the canyons.

I unpacked my few things and settled into a gentle existence. One of my most frequent climbing partners was soon Jim Yoder, who worked evenings at a lumber yard and spent days scoping, cleaning off and climbing new routes. The first time we met, he told me of an indignant letter to the editor that had just appeared in *Climbing* magazine, wherein a visitor decried the arrogance of Leavenworth locals. At Castle Rock, the visitor had asked some climbers for a guidebook. "I *am* the guidebook," one climber had replied. Fumed the visitor in his letter: "I'm not naming any names, but his initials are JY and he has the rattiest red beard in Leavenworth."

Jim Yoder looked straight at me as he got to this line. "JY," he said proudly, and pointed to his beard.

Halfway through the summer, Yoder hurt his arm at work. I called, inquired solicitously about his health, then asked if he'd mind if I poked around on his latest route project. He gave me his blessing and guaranteed that it was sure to be a classic, partly because it was 20 feet from the road. No sooner had I come back from the evening out (and some cleaning out, and some winging off) than the phone was ringing.

"Did you do it?"

"Um, yes."

"Shit," he said. Then immediately, generously: "Oh well, there's plenty of rock for everyone." I called the route Proteus, after my dad's sailboat, which had been named for the sea god who could change shape into an eagle or a dog or a fireball. (This seemed appropriate, since the climb required diverse techniques.) The boat's name had a double meaning, however, since my urologist father knew Proteus as a type of urinary bacteria. Plus the boat is yellow. My family's sense of humor is like that.

Another climbing partner was tall, wiry Jim Donini, who has so much rock talent he could come back after a long sales trip to Europe, a time of bourbon, food and fun, and climb disgustingly beautifully. Jim could talk and laugh a blue streak. Katie always said, "Put a quarter in him and watch him go." Donini certainly kept me on my toes. Ducking under a wire once, he roared, "AAAGH! It's electric!" as he jerked and convulsed. "*Jim!*" I shrieked. "Ha, ha," he said, walking on.

■ ■ ■

One of anyone's first places to climb in Leavenworth is Castle Rock, a solid gray fortress looming above the noisy Tumwater River canyon. One evening Katie and I set off on Gore Roof, which concludes with a foot-swinging hand traverse, then an overhang, above which I set up the anchors. Belaying, I felt the rope jerk sharply and thought Katie was taking a step back—I had, there at the start of the traverse, to re-think—so I paid out some slack. But she had just been dropping down onto her arms. After that, she fell, whooshed sideways and ended up 20 feet out in space. Kicking and waving, she tried to swim in to reach the rock, but just dangled. She shouted up to me, but because of the overhang, wind, river rapids, cars and trucks below, I couldn't make out one word. What I heard sounded like "peep-peep-PEEP-PEEP."

"What?"

"Peep-peep."

"WHAT?"

She paused between each word, gathering her strength to emphasize. "PEEP! PEEP! PEEP!"

Three motorcycles screamed by. It was getting dark. New to the cliff, I couldn't remember what rock features would be near her. If I lowered her to the end of the rope, would she even touch anything? Distractedly trying to recall how to escape from the system, I lurched around, tying off and adjusting. Then I stepped to the edge, and leaned out to where I could see Katie swinging gently, resignedly, in the breeze. I lowered her 30 feet to a stance, rappelled down to retrieve the protection beneath the roof, and climbed up again with a static belay. After that, Katie had to forge up via a completely different route, called Bone, into which someone had imaginatively stuck one.

Then there were the steep cracks and traverses on the gloaming Midnight Rock, in splendid solitude about a two-hour walk up a sandy path above Castle Rock. High on Midnight, scrabbling for jams in an offwidth named the Sting. I suddenly realized my head was stuck, I was about to fall, and I was going to hang myself. Stretch for that jam. No, wrong, resist temptation. Fish again. *OK*, do everything ve-ry slowly . . .

Once, when Neil Cannon visited, we heeded local (Yoder's) advice, and instead of hiking to Midnight Rock, drove up the ridge from behind. We parked the car and dropped on foot down and sideways through the woods. With no trail to follow, we overshot, but duly reached the cliff.

This scheme seemed all very smart until we got stalled on a climb, finished late, ended up hiking back in the dark, and overshot the parking pullout. Branches began sticking us in the eyeballs. We took one hesitant step at a time, dehydrated, hands blindly groping, falling with muffled shrieks off ten-foot boulders. It was 1:00 a.m. when we got back.

This, in retrospect, seemed pretty funny until our next trip when we did exactly the same thing. Warning: don't use this "shortcut" unless Jim Yoder is with you.

And down by the little A-frame there *was* bouldering—and by the boulders one day, a sunning rattlesnake on the warm rock at my foot. Just across the creek was Rat Creek Boulder, from which Katie and I would return with hands too tired to hold coffee cups. Among the myriad pools in the rushing Icicle River, in a place where I was led blindfolded, was the Secret Skinny Dipping Hole. On the rocks that ring it, boulder problems and a steep traverse make up Muscle Beach.

It didn't rain in Leavenworth. It just almost never did. The days blended together: sun, rocks, water, though I preferred idling by to swimming in a river named The Icicle. The A-frame was a peaceful place, with sun streaming in, a fat orange cat named Tilly, a pull-up bar in the bathroom doorway and a pegboard out back.

Aside from its pastries, town didn't divert us often. Leavenworth has no movie theater, for example. It was hard to find a reason not to climb. But there wasn't much reason to do so either. You'd wake up, make a pot of coffee, and maybe at noon, toprope some boulder problem and wonder, "How valid is this?"

On the other hand, you could go out and get truly thrashed with the tireless Jim Yoder. Once Neil and I made plans to climb after I'd finished my day's instructing. He went out with Jim in the morning. Mid-afternoon, I found Neil in the bakery where we drank coffee, ate bags of day-olds, and did crossword puzzles. "I can't climb," he said.

"What?"

"I have no more fingers," he said.

I was indignant. "*What?*"

"I can just hold the rope for you," he said. "Look, I have a note." He reached into his back pocket and retrieved a crumpled piece of paper scrawled with Yoder's handwriting: "Please excuse Neil from climbing today. He is not feeling well."

Leavenworth housed surprisingly few climbers for a town smack in the middle of the Cascades. But two were my boss Karl Schneider, Donini's go-getting director at Leavenworth Alpine Guides (LAG), and Dave Strutzman, another instructor. Dave, an Alaskan veteran, drove an old station wagon, wandered barefoot, and camped somewhere way up the canyon. He was tranquil about such things as bolting arguments. "People can do whatever they want," he said. One day when I took Karl to the gym, he told me he'd forsaken city life for the mountains. He'd had a wife, house, a picket fence, and was destined for corporate greatness. Karl told me of his plans for the LAG and, as he went tenaciously at his interminable workout, looked at me for feedback. I was at the bench press, slumped under the barbell. "I usually stop around this time," was all I could say.

Karl, Dave, and I were the only full-timers at LAG that summer. Within two years, both of those guys had died. Dave was buried, all but one foot, in an avalanche while he was skiing at Big Sky, Montana. Karl and a partner vanished into a crevasse on Mount Pisco in Peru. Their footprints ended at the scoured edge of an avalanche path.

■ ■ ■

It happened again recently—for one of my few vacation weeks, Katie talked me into a trip to Leavenworth. My first day there, Jeff Lowe and Mark Twight, nice and dirty from two days of putting up a new route on Mount Stewart, stopped in at Katie's A-frame. The Pacific Northwest has real mountain climbers. Imagine. It had been so long since I'd seen one.

Katie and I took a road trip up to Squamish, then came back down so she, having bought LAG, could attend to business. For my last day, Katie trudged me up a loose hillside to a "wild crack" she promised no one had ever done. A steep corner of thin edges led to push-palming and bridging up the seam. At the roof that caps the pitch, I ran into trouble—and the trouble with climbing with women. They're so goddamn nice to each other, so patient (I'm a sucker for it). I sunk awkward thumbs-down hand jams, moved my torso up with hands still low, and frogged my feet high. Caving in my ribs, I half-cornered the arête to get a foot up onto a shelf. Then I couldn't pull either hand out. I pumped out and rattled down to the rest. Repeating the process again and again and again, every time I panted, "Sorry . . . one more time," Katie would sing out cheerily, "Oh, no problem." So I ventured and retreated at least eight times. A gouge began to trench into one hand. But I really wanted to finish the climb and to name it the LAG Roof for Katie, Donini, Karl and Dave. Then came the magic moment when the jam felt just that tiny bit better. Visions of glory trumpeted; a great new route—maybe some copy so I could write off the trip! Then I was up, belly-flopping over the top.

Katie and I stumbled back down in the dark, clods rolling from under our feet. That night something woke me at 3:00 a.m.—electric pain in my elbows. In the morning, I tried to tell myself I'd been dreaming. But I thought, at least if I'd hurt myself, it was on something emotionally satisfying that tied the climbs and people of Leavenworth together. An epiphany, come to think of it. Then, the more I thought about it, the harder the route got. Pretty soon my internal dialogue declared, "11d, fer sher."

I called Yoder and asked him about that crack. He listened to my description, asked a question or two. "Yeah. Pumpline. 11a. I did it three years ago."

■ ■ ■

I make fun of the Leavenworth motif, calling the supermarket Der Safeway, scoffing at the ruffly aprons and the fact that the ringing that echoes across the valley each hour comes from speakers, not bells. But the bottom line is that the Leavenworth folks kept their home going.

Every time I talk to Katie, she says warmly, "There's always a place for you here. How about coming back?" I swear, that place is populated by magnets.

climb to safety
(in case of flash flood)

Pat Ament
January 1991

Pat Ament writes in this story: "How does one appraise his climbing, or his life?" One of the leading free climbers of the 1960s and early 1970s, Ament has devoted his writing career to such self-appraisal. For some, his stories are, as Ament quotes one letter, "Boulder's answer to Little Debbie's Peanut Butter Bars." For others, he is the greatest writer on climbing ever published. In this story, Ament sums up, offering a heartfelt and moving reminiscence about Yosemite and a life in climbing.

Outside the Tuolumne store I am greeted by John Bachar and several other climbers who are listening to a naughty rap music at tourist-scattering volume. A beautiful, cool air blows over and up the granite of Tuolumne. Sections of the rock where there is glacier polish are brown and sparkle. My companions, Steve Brawley and Tim Shultz, are eager to have rock-climbing challenges placed in front of them, climbs up domes with heart-jumping runouts above protection. Climbing to Steve and Tim, both 23, is urgent. When you are that age you are a god. You must do everything, and you can do everything. You will never die.

Bryce Wilson, not a great climber but very much of a character, joins us from L.A. We hoist him up what we can, a heap with a harness and a pair of sticky shoes. We don't care if we are held back, if the compensation is to be amused. The Bachar-Yerian route can wait—until another life. Bryce is tolerant of the indignities he suffers at our hands.

We find a campsite just east of Tioga Lake, add spices to some Dinty Moore stew, then stand out among alpine tufts in starlight. I recall my first visit to Yosemite at age 17 with Royal Robbins, our first night spent in a Tuolumne meadow where Royal pointed out constellations—Orion, Cassiopeia.... I still feel how much power radiated from that man, as he stirred from his sleeping bag the next dew-covered morning, ready to climb in his Valley. Who could forget Royal's heartless stare. In 1967 when I did the West Face of Sentinel in about 12 hours, going up two pitches in late afternoon, spending the night, and finishing the wall the next day with light left to have done those first two pitches, Royal stated that—because there had been a bivouac—it had to be considered a two-day ascent.

Steve, Tim, Bryce and I search sunlight for miraculous, small, granite knobs to which we guide our toes and fingers. The route offers a look outward, a sweep across the high country, or down to the river and the long, wide meadows, a vista that spreads in many directions. Sight is broader, more expansive, more in focus, than through any camera lens. You see with unbelievable depth of field, then in close at the bright, perfect stone as you locate a knob or as you notice the runout and feel a little rush of the willies. Years of your life pass by you as fast as that thrill.

Our plan is to drive back and forth between Tuolumne and Yosemite Valley, to play in both locations for about a week. I wanted to come back to Yosemite one more time, make a little review, have a few final reflections and return to favorite climbs—to once more run myself out up a wicked, vertical hand crack, and feel that type of fear, that sense of survival, which is unleashed sometimes almost in the form of a supernatural power. I wanted to drive through the 200-foot Sequoia, stare into the clear, dark, green pools of the Merced, study the greens of lichen, see old friends, bump into a few of them as casually as I always did. Would I chance upon any of those women I thought each different year were the answer to my problems, who rubbed lotion on my Tuolumne sunburns or let me play piano for them in the small chapel in Sentinel Meadow, or who I loved closely *in* Sentinel Meadow under stars?

Bryce has brought enough sleeping bags to stack into a Sealy Posturepedic. Tuolumne is disturbed by Tim's snore. In the middle of the night I hear Steve yell at him, "Shut up." And then more emphatically, "Shut up!" It's not easy to reconcile that such sounds should bray from one who climbs with so much sensitivity, who radiates almost a reverence toward the unfolding of a route.

Steve is a materialist—buys a car, a motorcycle, and, with his last fifty, a #3 Friend. A climbing store salesman tries to sell him a couple of artificial climbing holds, but Steve comes with us to collect real ones. He squeezes knobs as though each were the next delightful, irrational purchase.

Steve leads upward and anchors to an old, rotted sling. I correct the situation when he brings me up. I must monitor these children. We share a lone, golfball-size knob which is the largest in 400 feet.

Steve tells of when he worked as a waiter, how he took a woman's unfinished steak to wrap up for her, forgot, and instead threw it into the garbage—then remembered, retrieved it and gave it to her.

The four of us have breakfast one morning at the very small Tioga Lodge, where a ludicrous deer head is mounted on a wall and two forehooves are bent upward into L's for hat racks. The store sells postcards of redwoods, with paper probably made from redwoods. I overhear a tourist explain that the reason Ansel Adams was such a good photographer was that he had a special camera.

Driving into Yosemite Valley, I think of Howard Nemerov's line about trees, "To be a giant and keep quiet about it." There is a parade of RVs, the inhabitants of which peek out upon occasion to view Yosemite Falls or El Capitan, excited as they observe Texas Flake, which they believe has moved. What does it mean these days to climb El Capitan? Climbers throw their excrement off it packaged in paper bags, which you encounter as you stroll up along the base of the wall. People hangglide and parachute from El Cap. Paraplegics prusik it. It likely has been soloed by a two-headed woman. Royal Robbins would argue that this could not rightfully be considered a solo ascent!

There is a type of climber who *must* do the Shield on El Cap. They form a line at the foot of the wall. Unfeeling, they have no good connection with other people—like the climber in the '60s who, fearing a bust, swallowed his complete supply of acid and for years afterward struggled to make sense of what people were saying. These Shield climbers are quiet and speak only the language of the Shield. They file off the summit as perpetually as they lash themselves to the initial pitch. Rurp cracks up there are now bong cracks, three inches wide, a millimeter deep. Ting ting ting. You will always see a few tiny, red parkas high on the Shield. The people on the Shield never move. Rather they are replaced.

In the Camp 4 parking lot, a climber (just off the Shield) stands at his open VW bus door, hash pipe in one hand, beer in the other. He is in a trance, staring at a point on the ground, unbelieving of what he has climbed—or else he has been hit hard on the head by one of *those* bags.

I savor my July ascent of the Nose of El Cap in '67. It was the ninth or tenth ascent of the route. I was 19 and, to my parents' dismay, hocked my saxophone to afford the trip. My partner, Tom Ruwitch, was 17 and had only climbed a few months. El Cap was a wonderful rock school and provided a genuine dose of consciousness, probably worth a career in music.

I tour boulder problems I put up in Camp 4 in the late '60s after spaghetti dinners, on a night or two when the rock and I seemed divinely acquainted. The climbing world has no memory of these routes. Only spaghetti has survived. I walk around the problems, recall their footwork, their gymnastics, the strength and mind I gave them at the peak of my physical form, a soft-spoken Barry Bates sometimes bouldering with me. To touch these holds is poignant, taking me back to my obscure season of pushing extremes.

Brought to recollection also are strange freight-train wanderings to and from Yosemite—freight cars, starlight, alfalfa, or desert, when I carried within myself a new anticipation or fresh fix of Yosemite. All the years of climbing condense now into one acute, perplexing emotion I can show to the outside world only by being for a moment intensely quiet and looking away as though toward a sacred, distant place I alone see.

We walk slowly up the trail toward El Cap. Something swishes down into the trees—someone bungee-jumping from the Shield?

Leading the Sacherer Cracker without the right gear, spoiling for a fall, I think of its author, Frank Sacherer (killed in the Alps), of his bold leads during Yosemite's golden age, how he climbed quickly to conserve energy, onward, upward, often without protection, to the terror of his friends. I think of Chuck Pratt, who made lasting impressions on our minds with his mastery of cracks—of his and Royal's and Tom Frost's committing ascents, the walls of El Cap done without helicopters standing ready, without fixed ropes, without double-bolt rappel anchors leading down the wall, the quality of their adventures preserved.

Memory conjures obscure countenances—a fellow I spoke with in Yosemite in the '60s, who a day or two later drowned crossing a river in King's Canyon, climbers each with their own amazing spark of ability, who after a few months or years left Yosemite for good, who were too hurt by its changes or who did everything they desired to do, or the day of their inspiration was gone, or through drugs or loneliness they lost that extra will to care about climbing anymore.

One imagines Salathé and Steck up there during the first ascent of Sentinel. Harding, after a lot of wine, made plans with anyone for a new route on El Cap and then in the morning would not remember ever having talked to you! Mort Hempel playing his guitar among the trees. TM Herbert, in the visage of an Indian, talking about shooting squaws in the rumps with flaming arrows. Jim Bridwell transcending an impossible offwidth crack. The feeble, frightened whimpers of Chris Fredericks as he led a pitch. Pete Williamson hanging dead on his rope after drowning at the crest of Bridalveil Falls. Jim Madsen falling 3,000 feet from the summit of El Cap, trying to rappel in a storm with dry coats for Pratt and Fredericks. Layton Kor stemming wide, his pants split in two separate halves. Tom Frost's cheerful, truthful smile. Chouinard, living off tourist leftovers in the cafeteria of Yosemite Lodge and hammering Rurps into their early shapes. Mark Clemens, Pete Haan, Lauria, Breashears, Barber, Beck, Jardine, Erickson . . .

I recall a comment by my friend Tom Higgins, "There comes a time when climbing memories far outnumber climbing prospects." Yet his mentor Bob Kamps is leading 5.11 at age 60. The crack that now is eating my hands demands quick upward movement, like when Dave Rearick sat on a rattlesnake.

What a den of thieves and visionaries those times produced, some of us now priests and others cynical drunks who once were forebearers of a climbing magic and antecedent to the procession. We have not all heeded e. e. cummings's counsel:

You shall above all things be glad and young.
For if you're young, whatever life you wear
it will become you; and if you are glad
whatever's living will yourself become.

I think specifically of delicate, beautiful Shannon Wade. She is one of those giants who keeps quiet about it, but above all, she is young and glad. She looks freshly at climbing, certain that it is going somewhere, like her, budding—that climbing is not falling apart.

I discover that one of my haunts, the lounge of Yosemite Lodge, is no longer much of a hangout for chess players, musicians, and that smelly genus—the wet, tired climber. I recall playing the piano there, Breashears listening, and chess with Bachar. John's sunglasses had a tiny red light behind each lens that blinked on and off—a great tool of interchange when sitting among unsuspecting tourists. My gaze is diverted to a long, slender blonde who, upon closer study, I detect is a man. To be counted among the new order, it seems, one must be of dubious gender. Although I have never had hair on my face or my chest, I seem to be clearly a male.

All of these thoughts subtly interrupt the long sequence of the Sacherer Cracker. What this route needs is concentration on the texture of the rock rather than on how we compose and paint our lives. I hear the horns of a traffic jam along the road in the forest. I worry whether the young girl I have hired is looking after my cat in Colorado. For a moment the slightly dangerous runout seems ironic and to reflect the spiritual genocide of Yosemite. I envision myself falling, my head glancing fatally over the rock.

I think of that lone Bachar who has liked me and led me up a few climbs. At the top of New Dimensions, just where the crux of the route occurs, I was belayed by Bachar but took note of what it might feel to be there solo, as he had been, a thought which comes from what William James calls the "chill periphery." The largest challenge for Bachar is to keep such a spirit in any semblance of abeyance. The climbing of today's stars is so convincing that it requires courage to see them justly. Our previous story shall get some of its light from their vital chapters.

I see the names Pete Croft and Werner Braun on the Yosemite rescue list. How can people live in the Valley for such long periods and not die of its lack of culture? Is it that climbing and nature completely satisfy their needs? Somehow they must move about, in and around the surface view of Yosemite, and have an unusual intimacy with places no others go. Croft is a miracle upon the scene, with his calm traditionalism, an ability that shoots holes in Alan Watts' thesis that you *have* to hangdog and sport-siege in order to approach the new thresholds of skill.

Starting in the later '60s, climbing has been for me gradually less overt and more interior, less brute expression and more susceptibility. I learn that I know. Each technique is a memory experience. As a certain coincidence, hands find holds. Footwork has a charisma that is driving joy. I have 32 years of climbing. With half my former fitness, I better understand climbing's science of forces and benefit, as I succumb to obsolescence from a certain ascendancy (excuse the pun), a compulsion for establishing myself in improved positions of balance to reduce difficulty. Growing weaker inspires a new, different artistry. I try also to look more deeply at my friends and the physical environment than I could when I was so breezy, buoyant, and strong.

How does one appraise his climbing, or his life? For me it is as I appraise my writing. I am jeered and revered. A person from Wisconsin wrote that my prose is " . . . Boulder's answer to Little Debbie's Peanut Butter Bars" (a defense, or a critique?), while the anthologies of best climbing writings (*Games Climbers Play, Mirrors in the Cliffs*) display my work. The poet laureate of America, Howard Nemerov, recently wrote to me, of my poetry, "I am moved by your work, by its good feeling and its kindness." Yet some of my moments still have their hapax legomena. I didn't wait to be a polished writer before I exasperated people with the bulk of what anyone should ever say. Some see me walk down the street and think, "He must live in some clinic." Then I am invited to give a poetry reading at a university or be the guest speaker for the National British Mountaineering Conference.

My friend John Gill has told me to take pride in having an enigmatic persona. Because I live in me, I know who I am at my best. There is a tendency to assume other people should see that. Probably foremost of my faults is presumption. I have assumed also that the bad should not be seen.

Something Tom Frost has tried to teach me is to accept criticism of myself in good earnest. Life as it progresses takes on the nature of a mystical event, as though before you were half-conscious and now you are coming to. Those whose life work seemed to be to cut you down, turn out to appear reasonably discriminating. But most have tired of the effort, now that you are more open to it.

I see the text of my life as a very good one. It could have made a great difference for the better, perhaps, had there been a clearer destination in mind and had the images I pursued been less often at the stage—which ought to be preliminary—of impressions. One certainty: I have more friends that I love than I have people who hated me. Some of it is a love based on a faraway instant of days. Part of the pain of existence is that, even with good friends, there is a process of separation. Sometimes it happens quickly, or else slowly the while that you know each other. You go your ways, disassociate softly, for simple lack of time or as vicissitudes carry you to another place. Pratt would still recognize Roper and Royal. Does Schmitz ever think of Madsen? Where do I stand with Higgins, in light of his daughter and wife and work?

I think, it was because Royal loved us all that he was so hard on us. Then I think, no—he was latently egotistical and self-righteous, a grim reaper, bearing his dark plague. I return to the first thought. Royal, who Frost refers to in retrospect as "the Chief," truly was our spiritual leader, possessing our highest integrity.

Climbing is a lot easier than sustaining the better thought about anyone, but with Robbins the task is not so difficult.

Bryce must return home. Steve, Tim and I act as tourists, visit the gift shop for postcards and teddy bears. We take snapshots of high, towering granite. I must say to Galen Rowell that Yosemite is not a good test of one's photographic genius. Frankly, in Yosemite, any grandpa could raise his Instamatic into the air, click the shutter, and have a masterpiece. A few of those grandpas get their cameras slapped out of their hands by bears who have not given them a written release.

It snows the day before Higgins is to drive from Oakland to join us in Tuolumne. I phone and leave a message on his answering machine: "I guess you are reprieved of having to see me. After finishing the second pitch of Nerve Wrack Point the blackest cloud you have ever seen descended on us. Interesting way to end my final trip to Tuolumne. Farewell, my friend."

I have two parts of myself. One is a feeling that we will look back on this life as some trivial moment or some brief mission to gain knowledge. Another part of me, perhaps the atheist, conceives salvation to be the largest of our experience of life. While I try to look beyond life, I value every detail of what here may be all.

At dark a coyote ambles down the edge of a road. The coyote is soaked, for an instant confused, blinded by headlights, a moment of indecision and cold, rain dripping from its mouth. This wild, lost, lovely creature at last moves into the trees, stops, looks back, safe in the respirable green from people who move about with their laughter and contempt and their lights. There are times when you feel the pain of animals, how they struggle and suffer. You think that they didn't ask to be born, and that they deserve what dignity you can pay them. Somehow I feel that I asked to be born.

Mild flooding chases us to a rented tent in Curry Village, an expensive and lazy act. People with umbrellas tip-toe between deep puddles toward restrooms. John Muir and Teddy Roosevelt are among them. Steve and Tim write in their journals. They do not view me as the broken memory of a climber I feel I am becoming. To them I am one who has made climbing his help, a resource of stability, studies in how to hold on. I guess that's what climbing is, when you are going nowhere with it. It is something refined, something that keeps telling you that you are someone—when in your heart it is the tendency to believe that you are no one. It is a mystical knowledge that can't be told to anyone if they are not there yet. But what you can tell friends is each slant of your foot, and your respect for them.

Breakfast at Curry Village—then we take showers. The showers in Yosemite have never changed—they burn you and then freeze you, and then burn you. We drive by El Cap. Sunlight touches it in a way that reminds me there will always be a solitary, beautiful wall behind the circus. I notice a dark green pool in the Merced.

Tim says goodbye. He will head west to Santa Barbara to meet a woman. Tears which touch so many parts of me well visibly in his eyes, from having to part and to leave the soul-perplexing, colossal visuals of Yosemite cloud and rock. Steve and I begin our drive east into the sage of desert. I can feel the strangely moving power, a superior power, of those doorways of youth. I recall how climbing stole our hearts and was seen as a life that would be continuous with our own—sometimes absurdly, alarmingly hilarious. It provided those hallowed few instances when the very deepest meanings stood indelible to our minds.

Originally published under the title "Postscripts from the Edge."

confessions of a climbing instructor

John Long
March 1991

Weird clients and even weirder instructors are the subject of this John Long yarn. Fact or fiction? You never really know with Long's tales.

Max isn't so tall, but he's 200 pounds of corded muscle. If he were one ounce lighter I would have killed him years ago. Max is one of my closest friends—it's just that if Max were dead, I might never have to teach rock climbing again.

For ten years, Max has run a climbing school. On summer weekends, Max meets students in the parking lot of a restaurant, just past the heart of Idyllwild, California. There, in a covey of overheated cars, Max fits the students for rental boots, then collects waivers stating that if anyone should perish on the cliffside, it isn't Max's fault. That done, they carpool it up to Suicide or Tahquitz rock.

Fifteen clients had shown up for the "Beginning Rock Climbing Seminar," too many for Max and his two assistants to handle alone. So on this particular weekend, like so many others, I was working for Max. "Our guest instructor," Max said, introducing me to the clients. Guest, my ass. I needed the money.

We piled into cars and headed up to the crags. "Stay close together," Max had repeated about a dozen times. Yet by the time we arrived at the Suicide parking area, a straight shot of exactly one mile, we were missing the marine from Twenty-Nine Palms, a re-con sergeant with a mouthful of tobacco and the firebreathing Jeep. He'd taken the well-marked turnoff for Palm Springs, and it took Max 40 minutes to track him down. The beancounter from Brentwood was steamed. He'd paid his money. And not to hang out waiting for some jarhead, but to go rock climbing, for Christ's sake! He quickly worked out an equation in his head and put a dollar figure, relative to his day's fee, on the time we all spent waiting for Max to fetch the marine.

We hit the trail about 10:00 a.m. "Trail? Who said anything about a trail?!" balked the dentist from Pacoima. His teeth were too straight and too white for an honest man, and he didn't have a single hair on his bullet head. He did, however, have asthma. We would hike slowly, Max promised. Damn right we would. I had eight ropes and 90 pounds of gear in my pack, as did Max and the other two guides, Smith and Sandy.

Tall and coarse, Smith is his real name, mentioned here because I want the whole world to know what a fool he is. He's one of my closest friends, and he owes me hundreds of dollars. Sandy, an Australian, is Max's head guide. She has sandy hair, sandy eyes, sandy complexion—is altogether one of the sandiest women I've ever seen. All patience, she's a class act and a first-rate instructor. And she solos old aid routes with her eyes closed.

We queued up and started plodding up to Suicide, a steep, 45-minute march. I was angry at myself for having forgotten my hiking shoes, and having to hump it up there in a $100 pair of Air Jordans I'd just bought. What with the scree and gravel, my three-tone high-tops were trashed in ten minutes. My guiding fee was precisely the cost of another pair of Air Jordans, so I'd have to work the day just to break even. I locked my eyes onto the dental assistant in front of me—whose *svelte* loins bulged from her Nike shorts—and tried to forget about my new shoes. At steady intervals along the trail were rude puddles of blood-red tobacco juice, courtesy of the marine.

"Damn! Some poor sap decked-out soloing and he's crawled down," Smith started, "check out all the blood he's lost!" Smith proceeded to probe a lagoon of red juice with his finger. "The blood ain't even dry yet! Must have just happened. . . . "

"That's tobacco juice, "I told Smith, who was searching the surrounding thicket for a body. Smith looked puzzled for a moment, then stared me in the eye.

"I knew that," he said. We pushed on.

Just as we entered the pines and the trail hit the steep switchbacks, the asthmatic dentist started typhoon wheezing, snatching snorts on an inhaler between fits. Then suddenly, the girl in the Nike shorts held up, bent at the waist and went hands to knees, looking to wobble right over. The trail starts at 6,000 feet and most clients drive up from sea level the day they climb. The altitude and sudden exertion can do a job on them.

"You okay?" I asked. She turned her wan face toward me; her jaw dropped and she up-chucked a mean shaft of quiche and OJ directly onto my Air Jordans. Then her knees buckled and she swooned, tumbled into my arms and we pitched back, thumping to a halt in a welter of ropes and clanking gear, about three switchbacks down. I moved her over to a stump, pulled the community water jug from my pack and tried to wash the effluvium off my shoes. It took several gallons.

For about 20 minutes, the marine had entertained the front of the queue with mortifying tales of womanizing, oblivious to, or not caring about, the two ladies hiking directly behind him. Said women, both from North Hollywood, with close cropped hair, pendulous paste earrings, and matching terry cloth jumpsuits, exploded in unison.

"Listen up, Captain Pipe!" they bellowed. Max eased between them and somehow diffused things enough for us to carry on to the rest stop, where the fallen tree crosses the trail.

"Ee's gone," said Francisco, a Peruvian national who ran an import business out of La Puente.

"Gone?" sighed Max.

"Captain Pipe. Where ees Captain Pipe?" Francisco begged.

"Who cares?" carped the couple from North Hollywood. The Pacoima dentist sucked his inhaler dry and chucked the spent casing into the bushes.

"That man just threw rubbish onto the trail," said the grad student from UCLA. The dentist snarled at the student, then screwed another canister into his inhaler and huffed it hard. The Santa Monica triathlete, rigged out in black-spanned-cum-hot-pink, wrap-around shades, said Captain Pipe had "bolted" directly for the rock when the caravan had stopped for Francisco to take a leak. Max moaned and sent Sandy off to search for the Captain. She promised to try, but made no guarantees.

"Just follow the tobacco juice," said Smith. "Sorta like retracing the Nile. Bring boots."

"Shut up and take her pack," I told Smith.

"What are we standing around for?!" growled the Brentwood bean-counter. We plodded on, emerged from the trees and presently gained the base of the Weeping Wall.

Suddenly, there was a scream from above, followed by a long, blood-curdling "Fallllllling!!!" All eyes shot upward to behold a flailing body plummeting directly toward us. People dove clear as the body wrenched to a halt but ten feet overhead, thrashing like a swordfish on a gaff. Then he went limp.

"Lower me . . . ," the climber mumbled. Instead, the belayer simply unclasped the line and the dangling climber augered straight into the deck.

"I said lower me, not drop me, dumb shit!" the climber raged, quavering to his feet and rubbing several open wounds. His eyes suddenly beamed in recognition, and he went over and clasped Max's shoulders.

"Maximillian!"

"Hello, Danny," Max said quietly. Danny's wild eyes panned across the 14 horror-stricken clients. "Big class today, Max baby . . . Beginning seminar?" Max nodded. "Best time I ever had, taking your seminar,"

Danny said to no one in particular. "Hell, I took that class twice," he chuckled.

"Three times," Smith corrected.

"Dios mio!" Francisco mumbled.

"Well, what are we standing around for?!" whined the beancounter.

"Water!" the duo from North Hollywood demanded. I pulled the community water jug from my pack as Danny applied various salves and plasters to his weeping road burns.

"Didn't you even fill the damn thing up?!" pleaded the triathlete, eyeing the mere pint in the bottom of the three-gallon, plastic jug.

"I saw him pouring it on his shoes," the dink from UCLA chimed in, "like two gallons of it."

"On your shoes?!" gasped Smith.

Several valleys away, we heard the faint echo: "Captain Piiiiiiiipe!"

■ ■ ■

Due to hard work and clever promotion, Max has built his guide service into a tidy little business. Nothing remarkable about that. What is remarkable is that for ten years now, Max has taught climbing an average of 200 days a year. Most of the guides admit that, faced with the same prospect, they would have jumped off the summit of Suicide Rock long ago. But that's only what they say. The fact is, the job affords experiences, wacky and otherwise, that no other trade could hope to match.

And it is not true that the average student is a flake, a bumbler, or a demanding moron. Each class has its few "normal" people, inevitably the ones there for something more than a cheap thrill, the few who stick with the sport for the long haul. But never mind them. It is the really peculiar clientele that are most interesting. And nothing could have been stranger than the time we taught a beginning seminar for a gung-ho Orthodox Jewish temple.

Rabbi "Vincent" was the kind of youth leader you always wanted but never got. His tastes ran from scuba diving to skiing to parachuting and now to rock climbing, and he was good at them all. He'd flown transport jets for the Israeli army, had represented Israel in the Olympics (modern pentathlon), and was the kind of man you'd hide behind in a bar fight (not that he'd ever enter a bar). Max and Rabbi Vincent were fast friends. The two had climbed Tahquitz the previous month, and when Max had wrenched his ankle on the descent, Rabbi Vincent had carried Max out on his back. Anyway, as a 28-year-old associate rabbi, Vincent's duties included organizing and supervising all the weekend outings for the temple's youth group.

Max, Sandy, Smith and I met Rabbi Vincent, 26 wailing kids, and four other rabbis, at Quail Springs in Joshua Tree. It was about 100 degrees in the shade, if you could find any. The 16 boys wore gym clothes. The ten girls, in accordance with orthodox norms, could bare no flesh save their hands and face. They wore sweat suits under woolen skirts, plus long-sleeved shirts. That not one of them dropped from heat stroke, or converted to Christianity on the spot, says something about Jewish fortitude.

We were not allowed to actually touch the girls—also part of the orthodox code—so just tying them in was nearly impossible. But after an hour or so, we had six topropes strung over a gentle little slab and the kids were charging up the rock nonstop. Rabbi Vincent was nothing less than another guide, and his coaching helped get everyone up the slab. When things got out of hand, Rabbi Vincent would boom out a few words and the kids would stop stampeding over the ropes and chucking rocks at Smith.

At the flanks, under the meager shade of several Joshua trees, the four other rabbis looked on inscrutably. Three of these rabbis were turned out in the traditional black dress shoes, black slacks, black beard, black *yarmulke*, and long-sleeved, white dress shirt. The fourth rabbi, however, was unique in mien and form. Rabbi "Hank" was five-foot-six, and grossed out at about 450 pounds. And judging by the reverence shown him by kids and rabbis alike, he was apparently the high lama of the group, if not the whole temple. No sooner had we strung up all the ropes when the great rabbi took a stance by a flowering yucca, drew a small book of scripture from his vestments and began melodious chanting in a foreign tongue. Indeed, it was a strange tableau—the lone and level sands, the sun-tortured rocks, shrieking kids stampeding up the crags, and a chanting rabbi, leaning his bulk against a yucca tree.

After about an hour, when everyone had taken a lap up the rock, Rabbi Vincent called for a break and all the children gathered round him. It was then that Rabbi Hank stopped his chant and cleared his throat. The children instantly quieted and parted like the Red Sea as the mighty rabbi ambled toward the slab.

"Now," the rabbi started, "I will try." His keen eyes locked onto the slab. Then he spit into his fleshy palms and rubbed them together fast enough to start a fire. Rabbi Vincent's sunburnt face went white. Max reassured him that we'd get the boss up the slab, though he didn't know quite how.

"What?! We'll never get that tub of goo up."

"Smith, I will not allow this kind of language," I started, pulling him aside. "We are highly paid, professional guides, and we're here to do a job. A tough job, granted, but we will get the rabbi up the slab." Smith mumbled something about it being easier to raise Lazarus from the grave, but I paid him no mind. Rabbi Vincent was concerned, though. He told us that by hook or crook we'd somehow have to get the boss up the slab or Vincent would be skinning kosher chickens for the next month. Smith said he hoped Rabbi Vincent was fond of fowl.

Sandy scrambled topside to shore up the anchor, taking two dozen Friends and six Spectra runners for the task. As Smith and I puzzled over a directional ground anchor, Max and Rabbi Vincent tied Rabbi Hank in. Or tried to. . . . Our biggest harness was half again too small, and simply wrapping a jumbo-sized swami round the rabbi was no good. Owing to his pickle-barrel symmetry, there seemed no way to secure the bowline round his awesome paunch and make it stay there. Rabbi Vincent finally tied three, 15-foot swamis together, somehow lashed this loop round his arms, and fastened the loop through a colossal Swiss seat fashioned from yet another two swamis, and Rabbi Hank was ready to rock and roll. Frankly, all the peculiar rigging made the rabbi resemble a sort of porcine paratrooper, and the novelty of it all was not lost on the kids, who were sniggering and slapping each other's backs. Two of them were sent to the bus. Oblivious, Rabbi Hank was psyching hard, eyes fixed on the 40-degree slab as he huffed mightily through his nose.

The set-up was just this: The rope ran from Rabbi Hank and up through an anchor consisting of 17 assorted nuts and Friends, four bolts, and a ten-ton tied-off block. Then it went back down to the ground and through four locking krabs connected to three two-inch swamis girth-hitched round a stout, albeit stunted, Joshua tree. At the end of all this, with one foot braced against the stunted Joshua, I had the rabbi on a Munter hitch. Just behind me, Smith had me on a hip belay should the whole works somehow fail. But even Smith felt confident, what with the number of anchors employed, and the shrewd arrangement of same.

With Max and Rabbi Vincent soloing the slab on his right and left—for close tutelage—Rabbi Hank drew a last breath, the yuccas swayed and he started up. Amazingly, he was pretty agile, and sort of oozed uneventfully over the first 20 feet. About then I realized that something was all wrong, that it was all happening too easily. No sooner thought than several grommets burst from the rabbi's Fires like rivets on the Titanic. Then the laces blew out and the rabbi's boots were little more than sticky-soled moccasins, his plump feet ebbing from the leather as the stitching gave way. Panicked, Rabbi Hank suddenly yelled "Okay!" and immediately pushed away from the slab. Both Max and Rabbi Vincent lunged for him, but he shot beyond their grasp.

As 450 pounds plunged onto the cord, it instantly stretched to the merest piano wire. There must have been some slack out, because the topside anchors gave sounds of a head-on collision and my spine cracked as the rabbi shock-loaded onto my Munter hitch. All I recall is a colossal "pop!" then being dragged roughshod across the graveled base. When the dust settled, I looked up to see Rabbi Hank, unscathed and looking rather smug, staring down at Smith and me, lying in the dirt. Smith had his arms wrapped around my waist and was groaning horribly.

It seems that when the rabbi jumped off, the whole system had momentarily held fast. But the freight was simply too much for the stunted Joshua tree, which the line ran around, and which was consequently shorn off just above the ground. Then anchorless, Smith and I were perforce dragged into a line, which ripped the cord from Smith's hands; and as we bounced toward the slab, Smith, in perhaps his finest hour, had tackled me. Our combined weight, plus the friction of our bodies getting dragged across the desert, had slowed the great rabbi's fall, and he'd casually stepped to the ground.

Max and Rabbi Vincent dashed over to us and Max pulled me to my feet, leaving Smith writhing on the ground. Max slowly hiked up Smith's shirt to reveal a heinous rope burn, an 11-millimeter groove seared right down to his very gallbladder. Smith blared out that he was signing his check over to the P.L.O. before Max could slap a hand over Smith's mouth.

Shortly, a ranger showed up and took a hard look at the wee Joshua stump peeping from the ground. Then he went over and deliberated over the shorn tree, a host of swamis still girth-hitched round its splintered trunk. There would be a fine, of course, for the destruction of vegetable material. But Sandy broke down the whole scenario to the ranger and somehow talked him out of giving Max a ticket.

■ ■ ■

Every stinking time my bank account looks sweet, my wife plows her car into a Mercedes (down in Venezuela, where insurance is nonexistent), my daughter's tuition goes up ten-fold, or more likely than not, I invest in some sure-fire business deal, lose my huggers, and *voila*! I'm back teaching the water knot till I can shore things up. And the second I do, I swear never to teach another person to climb. This swearing business has become a bitter ritual with me, a self-imposed hex, really, because it is always instantly followed by me guiding for Max. Anyway, I cannot leave off my confessions without touching on some of the harebrained things that happen off the crags, but which directly relate to a guide's checkered life. One example . . .

I will call him Mr. X, and before he quit climbing five years ago—and quite suddenly at that—he was one of Max's steadiest private clients. Mr. X was a plain man. No, a homely man, really. Actually, I hadn't seen a face that hideous since the death masks in New Guinea. Over the years, all of us guides had led Mr. X up whatever climbs he had decided upon, typed out on a three-by-five file card, from which he tolerated no deviation. This slavery to itinerary was part of Mr. X's "goal orientation," a concept he was very high on. Aside from his goals, Mr. X was also uncommonly hale for 65, had a mean streak a mile wide, could be funny, and was absolutely weird in a way none of us could ever quite reckon until one of the guides—that man Smith—went to Mr. X's house for dinner. Now the maxim, "never fraternize with clients," is really a jewel of universal wisdom. If you doubt it, just listen to what happened to Smith.

Mr. X had a wife, call her "Trudy," in equal parts comely and strange, and no more than 35 years his junior. She was a painter, too distinguished to actually sell a canvas (I found her work incomprehensible), though she had "masterworks" in New York and Parisian museums, as she was quick to point out. Trudy would occasionally accompany Mr. X to the crags. That she considered climbing "asinine," and for "plebeians," she made no secret. This, and her relative youth, plus the fact that she seemed to have nothing in common with Mr. X whatsoever, made their union as confounding as Trudy's paintings.

More confusing still, is why Mr. X started requesting Smith as his private guide. Sure, Smith can climb 5.15d, but he is the very soul of irresponsibility, normally shows up an hour late, if at all, and his rows with clients are legend. Yet the two seemed to get on famously. Before that fateful night, I think Smith had guided Mr. X some two dozen times in the preceding few months. Though Mr. X had rarely tipped any of the other guides, Smith had bought himself a "new" used van with Mr. X's largesse. So you bet Smith accepted an invitation to dine with Mr. and Mrs. X.

Mr. X lived up in Holmby Hills, where a tool shed runs two million; and Mr. X didn't live in a tool shed. According to Smith, the first floor of Mr. X's pad was all shimmering sculptures, marble statues, and rare frescoes—a bona fide museum, except for a far wall befouled with "a bunch of Trudy's crap," as Smith put it.

They moved into the den. What with the lunar furnishings, the signed lithos, the dhurrie rugs and hand-painted china, Smith, who has even less culture than money, felt so awkward he wouldn't sit down until Mr. X shoved him into a chair. Immediately Mr. X began plying Smith with strong drink. Now Smith is a crag rat *par excellence*, though to my knowledge, he's never set boot to an actual mountain. But this has never checked his marvelous capacity for creating impossible facts regarding his alpine victories, then believing them himself so implicitly as to quickly win you over. So I imagine Smith had duped Mr. X up the North Face of Nuptse by the time they adjourned to the dining room and a candle-lit table set for three. Trudy wore a wispy saffron jumper that accentuated her mako curves and sent a tremor through Smith's legs. Trudy blushed, Mr. X cleared his throat, and they sat down.

And what an incredible meal—four courses, then cute little raspberry tarts and ice cream and cognac for Smith who was so full he unlatched his belt and had to flee for the bathroom to belch, flushing the toilet to mask the sound. Returning, Mr. X continued brimming Smith's glass till he was well-nigh blitzed.

Now obviously, I was not there that evening, and am going only on what Smith told me and the other guides. For what it's worth, Smith hasn't the imagination to invent the shocking events that follow; and if he was to lie, he'd deny any of it ever happened.

Briefly, Smith was petitioned to engage in a certain basic function with Trudy. I do not know who made the formal request, but at mention of the deed, the snifter froze in Smith's hand and he considered bolting without a reply. And by all the angels and saints, how he wished he had. But you certainly know about Smith by now, and he eventually agreed to the request, if begrudgingly. The three moved to the bedroom. I say three, because Mr. X was to observe. Anyway, perhaps ten minutes into the operation, Mr. X produced a laundry list concerning what duties he wished performed upon his beloved. It was then that Smith noticed that Mr. X, seated in a distant chair, his face a flushed and hideous mask, was fondling a 19th-century dueling pistol. Perhaps it was the sidearm, or Mr. X's slavery to the agenda, that set Smith to effecting the sordid list. Yet even Smith, a genetic fool further addled by liquor, had to draw the line on a certain act which he simply refused to name.

Now precisely what followed, no man will ever know, so savage did things become. Smith remembers only that a shot was fired, that Trudy booted him in the groin, that he was pistol-whipped "upside the bean," and that he escaped certain death through a glass door, after pitching a lamp through it. He claims he nearly blew his van up fleeing the scene, and I believe him. Yet the full drama had hardly begun.

Hearing the shot and shattering glass, alarmed neighbors rushed to their balconies to spot Smith's fleeing van. The authorities were notified. Shortly, Smith was nabbed near Sunset and Vine, and arrested for reckless driving. He was dead drunk, remember, which cost him his license and 400 bucks. But the real clincher was the fact that Smith was bucknaked. And what a fine time the coppers had putting the nude Smith through manifold sobriety tests in the middle of Hollywood's most notorious avenue, as bumper-to-bumper traffic hazed poor Smith from their Porsches and Jaguars. Smith swears that Zsa Zsa Gabor winked at him from her pink Bentley, but none of us believe that part. Anyway, Smith was eventually hauled off to the downtown pokey and, still naked, thrown into a drunk tank full of rummies and desperate rogues. And on that hangs another tale. . . .

largo's apprenticeship

Jim Bridwell
November 1991

No one has chronicled the adventures of Yosemite's climbers better than John Long, in tall tales hilarious and ribald. In this story, Jim Bridwell, himself frequently the subject of Long's yarns, turns the keyboard on the writer. In just a few hundred words, Bridwell captures the braggadocio, bullish strength and unflagging desire that made Long larger than life.

I knew them when they were children—Ron Kauk, Werner Braun, Lynn Hill, and many more. Well, they were children to me. Not that I was old, but I was older and I had been a regular fixture around the Yosemite scene for over ten years. Along with Mark Klemens, Jim Pettigrew and a few others, I had had dominion over bagging new routes, more or less at will, ever since my predecessors had bequeathed the kingdom to me in the late '60s.

Of the new generation, a seventeen-year-old John Long was the first to show up and make known his intentions to take the Valley by storm. One morning, Jim Donini strolled over to my campsite by the generator. He brought news of a recent arrival I had to meet. John Long was his name and climbing was his game. John was brash and outspoken, with a precocious appetite for the most difficult routes—of which he had a list.

This catalog of climbs was familiar. A product of Peter Hahn, it recorded only hard Yosemite routes done in the past two years. John entertained me with his contagious enthusiasm as I inspected the list. Immediately I could see the record of climbs offered complications for a Yosemite neophyte. Cream, Steppin' Out, Basket Case and other climbs on the scroll were cracks, involving the dreaded "offwidth" technique, a style of climbing nearly unique to Yosemite. I understood John was from the Los Angeles area, a Tahquitz regular and doubtless an excellent face climber.

"Have you done many offwidth cracks?"

"No," John blustered. "But they can't . . . and I can . . . "

"Hmmmm."

These offwidths were, at best, difficult to protect. Wishing to humble and not to harm, I offered to usher John up an easy, though exemplary, route: the left side of Reed's Pinnacle. John eagerly agreed and next morning we were tooling down the road toward Reed's in my '56 Ford. As we drove, John extolled his abilities, naming off the many testpiece face climbs he had mastered. I listened as I wheeled the old Ford through the series of turns, knowing that in all probability he would find this route somewhat different fare.

At the base, John announced his desire to lead the first pitch, a chimney of confining dimensions. Good enough—it would be hard to fall out, due to John's already sturdy stature. Off he charged like a bull at the cape. With a display of power, if not grace, John soon found himself at the belay. I followed using the practiced technique of a Valley regular, and quickly arrived at his side.

With little hesitation, I picked through the hardware and selected one wired nut (knowing the necessary size) and two carabiners, then started off. John looked bewildered, but said nothing, perhaps out of respect. I climbed up, clipped and moved past the bolt—purposely neglecting the rest spot. An interior crack on one side of the main fissure occasionally accepted the chosen nut. But then again, sometimes it didn't. This time it didn't, and the nut slid uninhibited down to the bolt. John's alarmed voice warned me while I moved through the crux section. I replied casually that I was aware of the fallen pro and that it didn't matter. Actually, I had soloed the route several times and felt solid, but certainly didn't want to let on to John and thus ruin the effect of my cool composure.

John attacked the crack. His muscles bulged and his veins popped. He neared the polished six-inch-wide crux section with little left but courage. Panic replaced what little offwidth technique he had. He powered on, prepared to die before he quit. (He tried to slump onto the rope for a cheater's rest, but I was having none of it and paid out slack in kind. If he made it up, I wanted him to know he had done it on his own.) His face was purple with effort and his once-powerful arms quivered. I finally took pity and divulged the secret rest hold, now behind his back. John's hand shot to it like a chameleon's tongue. Saved! Air flooded into his lungs in great vacuum-cleaner rushes. After a short rest, he swam his way to the top and my congratulations.

That evening at Camp 4, a friend, Phil Gleason, stopped by and suggested I have a try at a new route he'd been working on. As we talked, I could see the keen interest in John's eyes, so I suggested he might come along with Mark Klemens, my usual climbing partner, and me—if he wanted. Without hesitation, he grabbed at the chance.

The next morning—at sunrise—I was awakened by the drumming of John's pacing feet outside my tent. We threw some gear into a pack and went to the coffee shop. We were too lazy to make something ourselves, and it was free because the waitress lusted for me as I did for her. After breakfast, John still looked confused about the payment of the bill as we rode the shuttle bus toward our climb. The bus took us to the Ahwahnee Hotel, only a short walk from the route. Within a few minutes, we stood at the base of the described cliff. As expected, the flake arched above, leaning and overhanging. We drew stones and Klemens won the lead.

Mark's skills in flaring, overhanging offwidths were legendary, and John watched with awe at each precise movement. The flake leaned increasingly until the last eight feet, where it shot out horizontally. A young John Long sat next to me, totally confused as to a solution for this final bit. I'd been analyzing the problem since our arrival and had come up with the answer, but merely said to John, "You'll see."

Klemens showed incredible control while working hard to place protection. John fidgeted nervously just watching. Mark tried again and again to get something in before the crux. A bong would have been his preference, as he hated the newfangled nuts—as yet in their evolutionary infancy. He reached to the rack, selected a large I-beam-shaped contraption and announced with typical Klemens cynicism, "You know, I'm desperate now."

After using considerable energy fiddling with the piton, he threw it to the ground, cursing. Nearly spent, Mark finally managed to secure a bong, but without the strength to carry on, he lowered down to the deck.

My turn. Mark had set it up for me, having done all the hard work of putting in the pro. All I had to do was climb and clip. I climbed up to the high point at the crux. Then John made the mistake of looking away just as I slipped quickly through the tricky sequence. I'd gotten two reasonable fist jams, swung down and reached out to pinch the edge of the flake, pulled up into a lieback, and was resting before the boy from So Cal looked up again.

John was outspoken, to say the least, but only because he could back up his words with action. He started using the pure brute strength of his powerful arms, his feet bicycling. Through the echo chamber of the flake, I could hear his locomotive breathing. Once again he was desperate, but his great desire kept him afloat.

He'd thrashed and struggled to the crux, but now hadn't a clue. His life signs ebbed as I shouted down instructions which he followed to the letter. A hand flashed to the fin edge of the flake and his head and torso popped into view, gasping for air. A few power pulls and he'd done it.

"Good job, man!"

He would affectionately become known as Largo to his friends, and he and I would share many great adventures—the Nose in a day, the crossing of Borneo and others. But I'll never forget those first two days.

bad boy

Cameron M. Burns
January 1992

Climbing has its ugly sides, and few are uglier than the rock star who will do anything to become (and stay) famous. This wicked story is based on a real person, still an active climber, but Cam Burns isn't saying who. Anyway, half the fun is in the guessing.

Yeah, I climb. Twenty-nine years old and still crankin' hard. Got into climbing so I could nail babes. Dude, your lats are so cut, chicks can't say no. Besides, you talk a bunch of hardman jargon at 'em while you're wearing a tank top, and they get real giddy and flexible. There are megababes here in Fulton, but there are a whole lot of good climbers here, too. So it's hard to make a name for yourself.

I did it eight years ago. I became "The Bad Boy of Rock" by bringing in rap-bolting. Got tons and tons of shit from all the old guys. Shook 'em up a bit, but, hell, it was good for 'em. Good for me, too. I became infamous. The cover of *Rockclimber Magazine* had me, "Bob Thurmond, Bad Boy of Rock Climbing, on his route Brown Hamster . . . "

There's a competition at the Fulton Rock Gym. All the big names are here, plus a bunch of up-and-coming kids. The local media are all over it . . . same old shit . . .

Jim Merrill is giving an interview: "I climb because it focuses my attention on microholds and every little movement. . . . " Yakkety-yakkety-yakkety-yuck!" Who gives a living turd?

The newspaperman finishes up with Merrill and moves on to talk to Zoey Green, upshot little asshole. He's good, but he knows it, too.

"Zoey, I understand you are known as the bad boy of rock climbing."

Hey, what the??? I'm the Bad Boy . . . Zoey's just a little moron. All I know is I'm just glad I'm hittin' the road. A gravel enema sounds like more fun than hangin' round this geekfest. I can't wait to hit South Wall, California's latest bastion of Swiss-cheese rock.

South Wall is great. A few locals are developing the crag, but they suck. I'm going to redpoint one of their projects 'cause I'm way beyond these locals. I'll get the first ascent. I'm the Bad Boy.

A local Mc-fuckin'-yokel, comes up and asks me if he can take photos while I work the route. Whatever sends 'er, dude.

Tie in. Amble up to the crux sequence. Then it gets stout. Crimp . . . Foot . . . left . . . Crimp . . . Reach . . . Right foot high . . . Slap . . . Crimp . . . Pinch . . . Crimp . . . Fall!

Damn!!!

That last crimper is piss-useless and about as solid. Above it are two slopers, then the chains.

"Who bolted this thing?" I ask.

"Daryl Shamus. He's a pretty notorious guy," Local Dude says. "Last winter during a warm snap. He was callin' it Bad Boy."

"Bad Boy? What the hell? This is friggin' baloney, man. I'm the Bad Boy of Rock. Just pick up the goddamned May '87 issue of *Rockclimber Magazine*. It says it right there. I am the Bad Boy."

Local says nothing.

I tell him: "Yeah. Tell your friend Bad Breath that he fucked up that third clip. It's totally stupid. The guy must be an idiot."

McYokel shrugs like the friggin' McYokel he is.

I tell him: "You guys should learn to bolt these routes properly. Some of these lines are totally jingus."

Another shrug. Typical.

I try the route again but come off down low. Then I climb it out, falling again at the fourth pisser. This is really bogin' my high.

South Wall Canyon's an oven in the middle of the day. I hate sweating. I think I'll rest for the afternoon.

Later I run into Terry Bender as he's doing some pumps on a hard route. This guy has recently taken first place in a French competition. So he's probably the best free climber in the nation. Personally, I'd like to break his kneecaps . . . but I decide to say hello instead.

"Hey, Terry."

"Oh, hi, Bob." Terry's too polite to be a climber, let alone one of the best. He should go belay an aid climber for a week.

"You here for the State Nationals?"

"Yeah. How 'bout yourself?"

"Working a rig in the Garden."

"Good."

L. McY runs over with his camera and asks Terry if he can take some photos.

I tell McYokel, "Why don't you scram. Can't you see he's busy?"

"I don't mind, Bob." Terry turns to the camera jockey and says, "You can take some if you want."

"Thanks," McYokel says.

"Well, okay. But your fucking rappel rope better not get in Terry's way." I love treating the locals like dirt. You know, that's all it's really about anyway. Learning who to suck up to and who to treat like shit. Average Joe Lunchbucket thinks it's really about climbing.

I have to get back to Fulton tomorrow. It's either get the route now or someone else'll . . . snake it out from under me. This route has to go . . . NOW.

Three hours later it's dark and I'm back. There ain't hide nor hair of the so-called climbing fraternity now. (Ha! Fraternity's a good word for it. Bunch of juveniles with the cash for power tools and BMWs.) I head up the rock to where MY new route awaits. I've got my trusty Route Doctor in hand. A tap-tap here and a tap-tap there and HEY! I'm ready to do some climbing . . . Old MacDonald had Craftsman Tools. EIEIO.

Morning. I power down three espressos before launching out into South Wall Canyon. Cool, there are a few of the hardman contingency around. I'll just have to rub a little salt in.

I tie in and fire the route first try.

Local McYokel runs over and asks if I'll repeat it so he can take photos. Sure. What the hell? He can have a photo. I've just laid claim to this portion of the earth's surface. I own it . . . Ain't no one taking it anywhere . . . It's mine! I am the Bad Boy.

Local sets up a rappel and says, "kay."

I hop on and fire it again. The shutter blazes away a mile a minute somewhere above me. I need a sequined harness.

"How hard?" McYokel queries.

"Desperate. I changed the name, though." Not to mention the surface of the planet, but he'll never know . . .

"What's it called?"

"La Pieta. It's good. You should do it."

"Maybe."

I pack up my gear, tell the belayer "thanks" and blow. Gotta get home. On the way out of town I stop by the climbing store. Hell, better make sure this route gets into the history of the world. It's a delicate art, but I'm fully adept . . . .

After a few minutes of staring at drills, some sales dude cruises over to make a sale.

"Help you?"

"Naw, just lookin'."

Salesdude scopes my hands, the chalk and all the gobies.

"Been climbin'?"

"Yeah, just did some project up at South Wall. Guy named Shamus had bolted it."

"No way. You did Bad Boy!?!"

"Yeah, but I renamed it La Pieta."

Turns out Salesdude is the local correspondent for one of the mags. Cool. I tell him all about it. Make a bunch of all-encompassing-life-philosophy type statements so he can use a quote.

I tell him it's a 13d. It's probably only 12c, but 13d looks much better in print. Besides, sponsorship money means 13 or go for a walk. And, if you really must know, there's no such thing as a naturally occurring 13d in the world. NOWHERE! I defy you to find one. Why do you think Mr. Stanley invented chisels?

On my way out of the gear shop, I see a new poster of a Brit. The caption reads: "Bad Man of Rock Climbing, Freddy Bullen . . . "

Gaaaawd!!! What's the friggin' world coming to?!!

Back in Fulton, I'm getting ready for my next European trip. Then that magazine correspondent, Salesdude, phones. Yeah, like I didn't expect it . . . Not at all . . .

"Hey," he says. "Jamie Banger went up to try La Pieta and said that it had been chiseled."

"WHAT!!? No fuckin' way!!! What asshole did that?!!"

"I don't know, but Jamie said the route was 12a."

"WHAT!!?" I'm going to fucking kill someone!!! I can't believe that shit!!!"

"Well, I've been calling around and no one knows anything about it."

"Who's this Banger guy? Did he chip it?" Man, I'm pretty good at this. Look out, DeNiro, I smell an Oscar for the Bad Boy.

Salesdude says: "Naw, he'd never do that."

"Yeah, well you never know. People will do anything to get known."

I finish packing for Europe. Route Doctor goes into my pack, along with his two-inch-wide cousin Cliff Doctor, my homemade four-inch version dubbed The Mountain Mender, a bolt kit and forty hangers. I shouldn't bring any of it 'cause it weighs a ton, but what the hell? I'm out to change the face of climbing all over the world. I am the Bad Boy.

verve

Will Gadd
March 1993

Christian Griffith is among the most influential climbers of the last 20 years in the U.S. His introduction of sport climbing to Colorado, his writing about European climbing, and even his clothing line, Verve, helped change the way American climbers view themselves and practice the sport. More than a decade after his most influential climbs, he is still among the top ranks. Yet Griffith is also flamboyant and iconoclastic, and he is frequently the subject of parody—sometimes funny, sometimes cruel. In Will Gadd's insightful interview, Christian speaks for himself.

At 9:00 a.m. the Espresso Roma in Boulder is slowly filling with climbers and street types scoring their morning caffeine fixes. Although the temperature is barely above freezing, the sidewalk is steaming from the fierce sun. One climber, perched outside like a multicolored bird in a blue top, black vest, red tights, purse-like shoulder bag, black beret and multiple earrings, suddenly discards all the clothing on his upper body and sprawls out to let the winter sun soak into his skin. "Hi, CG," says a recent arrival and Christian Griffith extends one shaved arm (for better muscle definition) in an expansive greeting. Non-climbers take a few steps back and stare at him with an undisguised combination of curiosity and humor. Christian Griffith's day is underway.

At 9:30 a.m. Christian comments, "You know, the best thing about coffee is that if you drink enough of it you have to do something, even if you feel like doing nothing." With that caffeine-laden pronouncement, Christian is off to work at his company, Verve. Christian started Verve in 1987 with 30 chalkbags and a belief that he could produce better climbing accessories than any available. Verve has since grown into a quarter-million-dollar-a-year business, but Christian is still the sole owner and designer.

In addition to earnings from Verve and competition winnings, he collects fees for rock-gym design, course setting, writing and the odd speaking engagement. There is no separation between Christian, Incorporated and Christian—the public personality and the no-so-private individual. Christian is a master of self-promotion (coincidentally, his mother is a successful PR consultant) and is continually on the lookout for ways to improve or further his image. That Christian religiously practiced his signature to give it maximum appeal is a classic example of the carefully planned public persona he zealously pursues.

Christian's "office" is a small garage behind the house he recently purchased in Boulder. His work day is spent talking to his seamstresses, calling about orders and clearing a monumental backlog of messages on the Verve answering service. One message is from *Playboy* magazine—it is going to have a cover of a woman on a climbing wall, and wants Verve to supply the clothing. After consulting several female friends, Christian quickly makes up several provocative outfits with prominent logos and ships them out with the comment, "Can you imagine seeing a climber on the front of *Playboy?* It's too good an opportunity for Verve to pass up!"

At noon I visit Christian in the hopes of enticing him into a training session that evening, and make the mistake of wearing a sweatshirt made by another company, one that Christian dismisses with, "Just putting little trim around a baggy sweatshirt doesn't mean it has style. My clothing works for climbing; that sweatshirt works for posing." I laugh (the only appropriate response to much of his commentary), understanding that Christian's genuine creative drive is so strong that anything he didn't create is fundamentally flawed in his eyes. When the Paradise Climbing Gym opened in Denver, Christian critiqued its design with, "It's too plane-like, all right angles"—despite the fact that the gym is popular and successful.

Boulder is a town that abhors right angles as politically incorrect, and Christian is a born Boulderite. While he hasn't sold his mind to the prevalent New Age movement, he shuns most conventional medicine in favor of Pao, the Chinese Doctor. When Christian dislocated his shoulder at the 1990 Canadian Open, Pao had him doing push-ups just three days later. When he sprained his ankle, Pao drained about two cups of blood out of it. One climber commented, "Christian can pay me money to dance around in a grass skirt—it would have about the same effect!" The results of Christian's medical forays have been mixed—the shoulder has healed well, but the ankle is a source of serious irritation.

By four in the afternoon the temperature in Boulder has fallen to a crisp 45 degrees, just right for a fast bouldering trip up Flagstaff Mountain in Christian's latest toy and source of much joy, a newer used GTI (the last one was showing more signs of abuse than a piece of Soviet farm equipment). Christian spent $1500 having it lowered, $450 on tires and rims and topped the whole ensemble off with a $350 steering wheel. "This car is so much fun to drive that I will never succumb to the Dale Goddard mentality of reading all the *Consumer Reports* magazines and buying the most statistically reliable vehicle on the road. This car adds to my life!"

Most of Christian's friends agree, but think about their *own* lives when riding with him. Hilary Harris, Christian's housemate and bouldering partner, comments, "He's a good driver, but CG likes to live on the edge. He gets off by trying to scare the passenger—if you show fear, he just goes faster. I live with him, but I refuse to drive up Flag with him. That's the most terrifying thing in the world. When he retires from climbing I think he should become a race car driver."

Cyclists have literally leaped over the guard rail on the narrow, winding road to Flagstaff at the sight of Christian's GTI bearing down on them. When questioned about his driving, Christian gets defensive, maintaining, "Look, I've never had a bad accident and won't. When I'm driving like that, I am paying absolute attention. People have accidents when they *don't* pay attention."

On a one-day trip from Boulder to Rifle last fall, Christian and his car slid off the road and into a large hole when he "wasn't paying attention." Christian was irate that the hole was unmarked (as though it should be marked for people who slide completely off the road), angered by the change from pavement to gravel (there is a reason for the 15-mile-per-hour speed limit), and incensed that he did not beat his old speed record for the drive to Rifle. CG's anger is focused not on the fact that he slid off the road, but on the mistakes he feels other people made that contributed to his error. This extreme sense of self-confidence, often taken for arrogance, allows Christian to fly in the face of everything around him. When he introduced rap bolting into Eldorado Canyon and produced routes such as Paris Girl and Desdichado, he was condemned or ridiculed by many in the climbing community. In hindsight, those routes defined Eldorado Canyon as none have since Jules Verne, a runout 5.11 that Christian repeated at an early age. His names for Verve products (the Parisian, a chalkbag holder, is a euphemism for condom) often prompt sarcasm, but they are undoubtedly original.

It's 4:05, and Christian arrives at Flagstaff Mountain's sandstone boulders, which he has spent 17 years romancing. It was on Flagstaff that Christian first met Pat Ament, who reminisces about Christian with the genuine affection and pride of an old master watching his pupil come into his own: "Christian walked up Flagstaff almost every day to boulder because his mom had a house at the base. He was taken by my climbing ability, and I was taken by his spirit. I was astonished by his insights—for a fourteen-year-old boy—and how active his mind was. He noticed that I was slipping into oblivion in some ways in my life, growing older faster than I could keep up with. He tried to make me younger.

I was beyond my prime and going downhill, and he was moving at light speed toward inevitable notoriety.

"I'm not sure how this came about, maybe because of the creative approach his parents took in raising him. I taught him traditional ethics, and when he switched to rap-bolting it was almost a breach of trust, although I didn't treat it as such beyond taking a playful snipe at him in an article. Because I knew at the heart of CG was a great man and it was clouded by the glories of self-interest that take all of us occasionally. Christian has a vanity that he still flaunts, but I see him growing up now and coming to grips with some of the spiritual sides of life. If there was an imaginary competition for the all-time greatest boulderer on Flagstaff, Christian would probably get the award."

Flagstaff is the measuring stick by which Christian judges his climbing progress, especially before a competition. His pre-comp training ritual involves completing as many hard problems as he can in an hour or two, while wearing a weight belt. People tend to clear out of his path as he lopes from problem to problem, easily firing well-known *and* obscure desperates. As an important competition date approaches, Christian ups the difficulty of his circuit and adds weight to the belt, simultaneously reducing his caloric intake to the point where low blood sugar and fatigue can affect his performance.

By 7:00 the sun is setting, so Christian leaps back into his GTI and careens down the steep road in a driving performance as well-choreographed as any of the boulder problems he has just done. After breaking several dozen traffic rules, he pulls up in front of the Boulder Rock Club. The Rock was the second large gym Christian designed, after City Rock in Berkeley and before the Redmond Vertical Club near Seattle and Rockreation in Salt Lake. The design of each reflects Christian's continuing development, and all are commercial successes. John McGowan, owner of the Boulder Rock Club, comments, "He is a very creative individual, aside from all of his little quirks. When he gets working with geometric designs, some sort of brilliance shines through."

In the depths of the Rock Club lurks "The Igloo," one of the best-designated bouldering areas in the country, according to most who use it. Within its overhanging walls, a strange model of physics occurs as Christian and several other motivated boulderers orbit the space, repeatedly firing up the walls only to plummet back into the gravel landing pad. As I begin to doubt my own ability to do a move, Christian is there with an affirmative "GO ON!"—the right words to make me stick the last move of a problem that has spit me into the gravel so many times my hands look like cheap beef. According to Alan Lester, "Christian is one of the most genuinely encouraging climbing partners around. Some climbers say, 'Do it,' and don't really mean it, but Christian wholeheartedly wants you to succeed and knows how to bring the best out of you."

Christian has soon removed his shirt despite the Boulder Rock's attire rules, prompting several non-boulderers to look knowingly at each other and exchange remarks about "Lord Christian."

McGowan observes, "A lot of times he is arrogant to people, which can turn people off. But lots of intermediate climbers will get tips from Christian, and he goes out of his way to help people. There are a lot of people who hate him, but I often see him helping novice and aspiring climbers." Hilary Harris adds, "I'll defend Christian to the bitter end. He's one of the most awesome, sensitive human beings I've ever met. Most people who slander him recognize his accomplishments, but he's just such a good candidate because of the way he dresses and behaves."

At 8:30 Christian has been bouldering for four hours, outlasting several sets of training partners. It's time for food, another of his most passionate relationships. In one of Christian's early articles, he describes his continuing battle to stay light and ends by wondering, "How do you spell bulimia in French?" When I ask him if he has ever been bulimic, he says, "No, but have often wished I could be." At six-foot-two and 160 pounds, Christian won't win a heavyweight title anytime soon, but takes great pains to stay light—with the result that food is special, practically mystical or untouchable.

After dinner at The Creative Cafe (yes, this is Boulder), Christian attempts to entice several of us to watch a classic samurai film with him, but we decline, having sat through hours of stylized swordplay before. Christian and the samurai share a common culture, one in which style is essential to winning and self-absorption is manifested in ritually styled outward appearance.

The following conversations took place in coffeehouses across Boulder, driving to Rifle and suffering in isolation at various competitions.

How would you describe yourself?

Passionate to the point of obsessiveness, creative by necessity, driven and focused. Always searching for a purpose, a greater goal. Sensitive and insecure, but appreciative of the direction and attributes I have developed as a result of them.

How do you think others describe you?

I think people find me direct and decisive, yet it's often a fear of inner turmoil that makes me react quickly and strongly. People may also find me overly showy or vain. But I enjoy dressing and behaving differently, more for my own sake than for the attention I might gain from others. My sense of self is very plastic and it amuses me to change my outward appearance and speculate on similar inward transformation.

What weight do you place on public perceptions?

People's perceptions have never bothered me. I would like to think that most people think well of me, but there are very few tangible (social) things that people could find fault with that would bother me. I have always felt a bit outcast and I don't find it threatening—quite often I crave its privacy.

You are often accused of being somewhat of an exhibitionist. For example, taking your shirt off at various competitions.

Do I leave my shirt off at those events?

Yeah, any comment?

I never liked wearing shirts. When I was in Joshua Tree I used to get up before anyone else, go about a hundred yards outside camp, strip off all my clothes, take off my shoes and go running out across the desert. I'd come back in an hour or two, having run for miles across that beautiful desert with no one around. For me it's never been exhibitionistic, but more a level of comfort that I prefer. That's what my whole clothing line has been about. If you are going to have to wear clothing—and I'd rather not—then it should complement you and be incredibly functional, not for the sake of fashion or style. That's why Verve clothing is only made for climbers. Because I don't want to compromise it. I'm a climber, and it would be crazy of me to make things that weren't exactly what I needed.

I remember walking out to Equinox one time, and I was just so into the whole environment that I stripped off all my clothes to enjoy it. Alan Lester was so embarrassed—he's fully clothed, and I'm walking along nude in a pair of sandals!

I used to solo in my underwear in Eldo all the time. Maybe there is some subconscious base to what people consider my exhibitionism, but it's really just the joy of nature, and I do feel very comfortable in my body most of the time. Why cover it up unless you have to?

I don't know how aware you are of it, but you do attract a fair amount of sarcasm.

You mean people making fun of me?

Yeah, your friends are totally loyal, but people do like to make you a target.

It's funny because I'm not really aware of that dichotomy. I've heard that I seem arrogant occasionally—or intimidating. I can see people thinking that I am focused and intense, but arrogant doesn't quite seem right. Intimidating doesn't either. A lot of people misinterpret intensity for arrogance. When I'm focused, I'm not willing or able to match another person's level of intensity.

When I'm feeling good or right, I don't really hold back—I express myself. There are times when it's important to let the spirit of the moment have free reign. Sometimes I do it very directly and bluntly; or I'm carefree, although someone could take it as being self-centered. A lot of this comes when the moment is on you, and I live for that feeling. The consequences aren't necessarily important.

When I wrote the "Manifesto" [see *Climbing* #100], I really felt like there was a whole side of the sport that was under attack from people like John Bachar, Ron Kauk and Lynn Hill and all those people who are now sport climbing! But they were saying that this was the biggest atrocity in climbing and that it had no validity. I wrote the "Manifesto" in a car on the way back from Smith Rock as a response to their attacks.

How do you respond to criticism?

Some of it, like that paper doll thing in "The World According to Griffith" [*Rock & Ice* #14], was really funny. Some of it isn't, but that's part of the game and I set myself up for it. A little good humor can make you see yourself more realistically. I have certainly been bold and grandiose at times, and far from hidden and unassuming. For example, I wore tights at the American Alpine Club for my presentation so I could do the splits on stage. That could be considered exhibitionistic, but everyone stayed awake for my slide show. It was lively, good entertainment, and that's what I was there to do. People could say, "God I can't believe Christian wore tights up there." It was fun and something I could do, and so I thought I would show it to everyone. It's weird now because I've got a business and feel much more limited in what I can do.

It doesn't seem to have limited you too much.

Yeah, I just named a route "The Gay Science."

What's the story behind that?

It's a book by Nietzsche. It's his ethics, a way of living. And it also sounds kind of racy, and will turn a lot of eyes. Besides, it pokes fun at all the macho route names like Poetic Justice, The Beast, Colinator, which seem so prevalent at Rifle. Then I thought about it and wasn't sure I should have named it that. Still, I imagine there aren't any other Gay Sciences in the world. It's always been important for me to be unique.

You have competed extensively and well in competitions. Are they ultimately good for the sport? For climbers?

Climbing competitions are good for the sport and they are a logical extension of its development. They are wonderful social venues and have brought the American scene to a sort of solidarity that it has never before known. However, it's easy when preparing for competitions to forget the fact that one is a climber and simply become someone who trains. I like the physical directness of training, but the psychological/spiritual aspects of the sport, the dance and personal connection with the rock, get neglected. Maybe that's why so many climbers burn out after long comp tours.

Have you ever done any mountaineering?

When I was young I climbed a fourteener [14,000-foot peak] with my father almost every summer weekend. But except for a few grade V (mostly rock) routes in Chamonix, I have done none. Though I did free-solo the ridge route on the Matterhorn in running shoes and tights in record time (literally!) back in 1986.

What do you regard as your important climbs?

To start with, Zombies on the Lookout, an obscure 5.9+ in Eldorado. I led the first ascent when I was 15. It was rotten, licheny and had no gear for 100 feet. When I got to the top, I was so mentally blown I wasn't sure if I had fallen off and died or not. When Pat Ament finally reached the belay, I was so relieved, because only then was I certain that I was tied into the belay and not lying down in the talus somewhere.

The Rainbow Wall. And Paris Girl is obviously important. It was probably the first 8a in Eldorado and was also the first route rap bolts were put in on. When it got chopped and I replaced the bolts, things really got heated. I knew I was right about the direction the sport was going, and I loved being a protagonist of change. My letters to the editor and "Manifesto" were all inspired during this time.

I shouldn't neglect Desdichado ("Disinherited"). It's such a brilliant sequence with such great position. When I did it in 1985, though I didn't realize it, it was the hardest route in the country. Verve is another, as are the Red Dihedral and the Bishop Crack.

When will you stop climbing?

I hope I will never stop. Climbing has made me who I am. While I may not be able to climb the most difficult grades at some point, becoming a master of the sport is what I'm really interested in, and that's a much longer process that ultimately transcends physical limitations. I can remember showing [Patrick] Edlinger around the Boulder Rock Club and watching him climb a ladder to the top of the bouldering area. As he stepped up off the last rung, the simplest of climbing movements, he was in such perfect, yet unselfconscious, control that I was amazed. That sort of mastery takes decades to attain, not just a couple of years of hard training.

You graduated from the University of Colorado at Boulder. Why did you do a dual major in philosophy and psychology?

I have always felt strongly the need to ask questions like, "Why am I here? Why do I feel this way? What is really important?" I don't think I learned any answers at school, but I did learn how to better ask questions. I am also curious about the human psyche, what drives it and what it does under pressure. I have always felt a certain affinity with people who have brought extreme situations upon themselves. One of the authors I really like, Jean Genet, was a sadistic homosexual who basically spent all of his life in prison and wrote his books on toilet paper. Still, his experiences shine through the depravity he was existing in. He wrote this stuff for what purpose and reason? He was just down there being human in the midst of this internal and external turmoil. How people have resolved those conflicts and endured to make the rest of their lives meaningful is fascinating.

There was a long time where I didn't think that going out and doing something because it was "fun" had any value. Everything had to be really directed toward some greater goal, some inner development. "Fun" used to not exist except as a distraction. You don't climb professionally to support a habit of going to amusement parks. It's an inner sense of purpose that drives you and makes anything else seem foolish and childlike.

What part of climbing do you value most?

I love the pure focus of climbing. The release and yet connection that one achieves only on rare occasions (for me, less than seven times in my career) when extreme concentration unlocks an intuitive presence in which the climber is neither in his/her body, nor out of it. It is a moment of unreflected action that has a divine quality.

Is climbing healthy, or an addiction like crack?

Often you see older climbers that were once very good, and they seem worn out or strange. It's easy to say, "Look at what climbing did to that poor sucker." But where would these people be *without* climbing? At the very least, didn't they have a moment in the sun? I was once told an athlete dies twice—when he quits his sport and when he dies physically. I think this is true. It is hard to find situations in modern life that have the same intensity and sense of indelible purpose. To lose that is truly a death, yet this doesn't devalue the extreme, it only defines it. For myself, the choice has always been clear.

When I was in college, I used to hope the world would end before I graduated. I wasn't depressed—just so focused on my climbing that I couldn't imagine any other kind of life. I had the feeling that when college ended, the bubble would burst, and I would be forced to deal with the fact that all I had been was a climber. What would I do for work? These questions seemed overwhelming. If the world had ended, so much the better for me. I would have lived my life doing exactly what I wanted.

If you died tomorrow, how would you like to be remembered?

I do not want to be remembered as someone who simply did things or took things without giving back. I owe most, if not all, of who I am to climbing. I hope that I have been successful in repaying the efforts of people that helped introduce the sport to me. I have tried to use my creativity and my passion to rekindle in climbing these aspects which it has inspired in me.

You were a protégé of Pat Ament when you were younger. What was the nature of that relationship?

Pat imbued in me that rock climbing can become more than just a hobby, toward a spiritual way of developing yourself. If you took it seriously, it could become a beacon in your life that could direct you.

I met him when I was 12 or 13, when my friends were starting to experiment with alcohol and other drugs. Pat suggested that I was lucky never to have tried these drugs, because they could ultimately destroy my kinesthetic edge. If you want to become a great climber, you must protect that edge. Consequently, I was able to refuse John Sherman's beer bongs at the age of 15.

Pat also gave me my first real taste of publicity in an article he wrote entitled, "Piecing Together Split Rocks."

How important is a sense of history to climbing?

In the old days, climbers like Jerry Moffatt, and even me to a small extent, had a strong background in really dangerous and bold ascents. Jerry did things that were cutting edge when he did them, real milestones in British climbing. He wasn't just a competition climber, someone who had really strong fingers and worked out all the time. A certain amount of his whole persona is wrapped up in the rock; he's not just a skilled puppet. I think that sense of history allows people to maintain a little personal integrity in the face of these really social circumstances like climbing gyms.

Do you think chipping and gluing are the future of the sport?

It's pretty obvious that both those things are going to play some sort of role in the future of the sport, and you would be setting yourself up to be proven wrong or be seen as a hypocrite if you said otherwise. I was saying years ago that the hardest routes of the future would be manmade. The closer you get to the line between possible and impossible at a given standard, the fewer your possibilities for actually being able to find it.

I hesitate because gluing and chipping are so devastating in the wrong hands. It's like rap-bolting; back when I started doing that, you only rap-bolted to establish routes that were of the par that set that precedent. On 5.13a and above, rap-bolts were OK; 12c and below, hooks were good because the holds were big enough and the angles not so steep. Maybe chipping and gluing fits on things that are hard 5.13 and above. Which it seems like people are doing, whether or not they are calling it that.

La Rose et la Vampire [at Buoux]?
    Yeah, La Rose, and I think Hubble has a lot of glue on it.
It seems like many good new routes do.
    You are talking about such small amounts of stone that it makes sense to protect it to some extent. Breaking off a tiny fragment of rock—something the size of a fingernail clipping—can make an extreme route impossible.
Are you starting to split hairs about what is OK?
    I think it's obvious that people have committed heinous crimes against the rock, using bolts, glue and chisels. I don't think that, as a course setter who has set dozens of routes for competitions, I would feel comfortable just going up and carving something out of the side of a cliff. And I don't think there are a lot of climbers out there who have the experience to do something like that. But people seem to think they do and they just do it, and it's gross.
Do you think La Rose is gross?
    La Rose is a mixed route. Some of the moves are beautiful. Antoine [LeMenestrel], who chipped it, is one of the master course setters. Accroche-Coeur is a more gruesome example.
    It probably has a lot to do with the area where you are. You could create some amazing boulder problems on some of the boulders in Yosemite, but is it worth it?
Have you created routes to raise the standards, such as Verve in Boulder Canyon?
    I don't know who smoothed out those holds, but a few definitely shouldn't be there. I used to always say that if you could do the move on a hold once, no matter how shaky the hold was, then you can glue it, because it allowed one ascent and was therefore a viable hold. It's not like you tenderly go up and pry off this big flake, sculpt a bunch of Sika on the back and then stick it back on. I have never done that.
You originally rated Lakme 14a, and received a great deal of press for it. It turned out to be much easier.
    I didn't receive as much press as I *should* have if it really did turn out to be 14a, considering what Scott Franklin got for Scarface!
    I've probably been guilty of over-grading—certainly on Lakme. People tend to put so much effort into the sport that they hold themselves in too high a regard and don't question their own fallibilities. There is also so much pressure not to be left behind, that I think a lot of people artificially boost themselves up, to the point where they say they onsighted or flashed something.

We've had some long talks about 'days' versus 'tries' for quantifying relative effort on a route.

People are straining the envelope and using the loosest possible definition of an ascent. What do you mean by onsight or flash? "Well, I tried it last year but I had forgotten everything, so I really feel like I flashed it." You didn't.

I think people tend to be really fixated on numbers. But it's the time when you feel the flow and unity of your body and mind and rock—and whatever you are trying to do—that is really valuable. Not going back to the car, saying, "Yeah, I ticked so many hard routes today."

paying for the summit

Jonathan Waterman
March 1994

The guide's life is one that every serious climber must consider. Climbing all the time, and getting paid for it—what could be better? Jonathan Waterman gives us an inkling of guiding's occupational and mental hazards in this story of his last professional climb of North America's highest peak.

In the spring of 1981 I signed on under Michael Covington as a mountain guide on Denali. My boss wore an electrified afro, a euphoric grin and a Nepalese Z-stone necklace. Michael insisted that he loved guiding; he'd sooner rot, he said, than work nine to five. Despite the pollution surrounding Michael's ever-lit "Slim Sherman" cigars, the director of Fantasy Ridge Alpinism became my role model. After hiring me, he provided plane fare and a $700 monthly wage. Even though seasoned Denali guides made twice as much, most climbers didn't grouse about free climbing trips. Furthermore, the salary included room and board—a tent plus all the freeze-dried food you could eat.

After climbing the West Rib with John Thackray for fun, I arrived back at basecamp. Here I fell into repose and awaited my job, guiding 18 circuitous miles to Denali's 20,320-foot summit. I assured Michael that reclimbing the mountain with clients would be a real cruise.

Guide mercenaries can be identified by white "raccoon" eye rings where sunglasses have blocked the beginnings of skin cancer elsewhere on their faces. They also share a cologne of urine and sweat mixed into their synthetic clothing. Some of these guides are unassuming and soft-spoken. Most curse, chew, spit, smoke, fart, brag, blow harmonicas, sing off key, flaunt their egos and conduct themselves in a manner that would lead to their excommunication from church, their mortification in public or their eviction from apartments.

I was thus not surprised when my fellow guide, Steve Gall, sauntered out of the skiplane wearing a bandana, pirate-style, around his forehead. He was blowing smoke rings from a Slim Sherman wedged between his fingers.

As the seven clients jumped (and fell) out of the skiplane, I began to envision our group dynamics. Even a neophyte Denali guide like me could sense that there were insurmountable obstacles ahead. Our congenitally uncoordinated Baptist minister, Duane, told us that we didn't have to worry because God would see us safely up the mountain. Duane had been directly appointed by his own Chief Guide in the Sky to climb the mountain so he could return home and better direct his misguided flock, sinners all, from some Baptist backwater of southern Georgia. Our minister was, according to Denali guide parlance, only one of "a school of tuna."

Our biggest catches were the Texans, Ernest and Evelyn, living out the great American dream by climbing America's tallest peak. Ernest was 40 pounds overweight with a heart condition, while his pale, slender wife would easily kite away if the wind blew over 50 miles per hour. In this business, however, the consummate Denali guide knows that as long as his client's money is green, anyone can pay for the summit.

My colleague Steve seemed strangely resigned to his new charges. But our troubles began only a mile up the Kahiltna Glacier, 17 long miles from the summit. The Texans, with the exception of their 19-year-old son Walt, could not drag their sleds; Steve took Evelyn's and I took Ernest's. Another mile farther, Steve lashed Evelyn's pack on top of his. A hundred yards from camp, Ernest collapsed.

He clutched his chest as his pulse raced like the traffic on Interstate 20. When he calmed down an hour later, he allowed that he would have been fine if he had taken his heart medicine.

While slumped in his sleeping bag, Ernest summoned his most authoritative basso and announced that he would be continuing up the mountain in the morning. I agreed, in order to comfort him, then continued plying him with tea as an excuse to monitor his speedy pulse and erratic breathing. At 7:00 a.m., after quietly radioing Doug Geeting for an evacuation, I bulldozed the protesting Ernest into the skiplane before his wife could interfere. I told Geeting not to bring Ernest back to the mountain under any circumstances, even though he had paid the $1200 guide fee (which Michael later refunded).

That afternoon, while Steve shuttled loads with the others, I escorted Evelyn back to the landing strip so she could join Ernest in Talkeetna. After I lifted Evelyn aboard the plane, I asked the bushpilot, Don Lee, if he'd given my note to my friend Chris Kerrebrock, who was attempting the Wickersham Wall on the other side of the mountain. Don's face dropped.

"You didn't hear?" he asked.

"No."

Don walked me to the side of the airstrip. "Chris died in a crevasse fall."

I thanked Don for telling me, walked over the hill and out of sight of the climbers at basecamp, then screamed as loud as I could. "Why him?" I yelled. Suddenly climbing Denali seemed so frivolous and self-indulgent that I seriously considered commandeering the next plane to Talkeetna.

Certainly, no climber is immune to the death of a friend. It can happen to anyone at anytime, but when it does, the surprise is inconceivable. The sense of loss darkens the dazzling light of the mountain; it sours the sweetest camaraderie; it turns all reverence to disdain. I felt (as the late Lionel Terray had suggested in the title of his climbing memoirs) that I had become a Conquistador of the Useless.

It would be difficult to resume climbing now, even with a friend, let alone five paying clients who had never been on a big mountain. Escorting charges up the West Buttress was going to take everything I had. I had to convince myself that, since Chris had been a guide, he would've stayed with his clients in a similar situation. I began skiing the five miles back up the glacier.

When I came to the first easily crossed crevasse and stared down into its onyx throat, I was so stricken that I had to sit down. I imagined what it would be like to be wedged 50 feet down, as Chris had been, unhurt but hopelessly stuck. Chris knew he was finished, so he asked his companion, Jim Wickwire, to relay messages of love to friends and family. Chris thanked Jim for his help and made him promise to wait for a partner before continuing up the glacier. Then Chris summoned his courage and waited for the lying warmth of hypothermia to take him away.

I cursed the crevasse until my throat grew hoarse, then steeled myself. As I glided over the narrow crack, I imagined free-falling into black space—like Theodore Koven, Allen Carpé, Jacques Batkin, Johnny Mallon Waterman and Chris Kerrebrock all had done on Denali. My heart was pounding in arrhythmic terror.

Meanwhile, Steve was dealing with our sanctimonious client Duane, who tripped over his snowshoes, dragged on the end of the rope and spilled a pot of water in the tent. When I finally returned to the 8,000-foot camp in a blue funk, I heard a tirade.

"Duane," Steve shouted, "you're nothing but a low-life, worm-eating, shit-licking, son of a whore!"

There was nothing to do but laugh. I was back in the fray. I had to concentrate on the clients' needs—and grieve about Chris while alone in my tent.

Our Baptist minister did not take kindly to Steve's rebukes. After a few days, following every new assault of profanity, Duane gently chided,

"Stevey, I jus' doan think y'all can talk to a human bean' this way."

Over the coming weeks, the other clients—Jack, Fran, Bruce and young Walt—found solace in Steve's latest invective against Duane. They would titter at every new "maggot" or "dickhead" or "pussyface" as if Steve was lampooning the Ayatollah. It became apparent that Steve was trying to relieve everyone from the frustration of Duane. Diurnally, Duane would spin judgmental Bible lessons to anyone within earshot; nocturnally, he would bellow out a cacophony of snoring beside his tentmates.

When sharing a tent with me one night, he implored, "Jonny, y'all really need to spend some time with this good book here." He knew that I was mourning Chris; Duane wanted to help. He would lift the Bible and touch my shoulder when he talked, and I let him sermonize unhindered, because a good guide is supposed to listen—despite Duane's drooling lower lip and basset-hound face, despite his righteous and windy lectures.

After the first week, Steve and I agreed that we could not take Duane to the summit because he would endanger the team if he ran out of gas the way he did while carrying loads. So Steve and I alternated babysitting Duane and keeping him on a tight rope.

■ ■ ■

Statistically speaking, Duane was fairly safe. During the 1980s, out of 2,284 clients who attempted the summit, only one died; out of 5,247 nonguided climbers, 33 died (five in crevasses). Most Denali guides will concede that every team of clients is stacked with a walking time bomb like Duane. Usually, the Duanes retreat or get evacuated after the first few days. The fact that they rarely die, fall in crevasses, break legs, frostbite toes, scald fingers, torch hairdos, asphyxiate tentmates or are strangled by their guides has a lot to do with the incessant vigilance of those guides. Duane, however, was convinced it had more to do with miracles.

While shuttling loads to 14,300 feet, I warned Duane to step over the obvious crevasse; I was particularly concerned about crevasses lately. (By this time most of the climbers on the mountain had heard that four rangers took an entire day, chopping ice and rigging pulleys, to remove Chris's corpse from the crevasse.) Duane paused at the edge of a 16-inch-by-200-foot slot of blue space, squinted his eyes as if to make a calculation, then stepped directly into the hole. In the 10 seconds it took me to sprint back to him, he slipped into the slot like a greased pig into a hay shredder, kicking away and widening the crevasse walls. He stretched the rope leading to a teammate and sunk in to his hips, then up to his chest. When I arrived at the lip, he was undermining his last vestige of support. I grabbed his chest harness and wrestled him back onto the glacier. Duane flopped on the snow with sweat streaming down his face, lips funneling for air, carabiners jangling on his harness. When he finally spoke, he thanked God for his salvation. I was not mentioned.

That night Steve and I pitched our own tent 50 yards away and out of earshot of Duane. The others gave Duane a gag order: All sermons except emergency prayers were forbidden. We had already put knives, matches, stoves and cooking off limits to Duane.

■ ■ ■

Duane had pushed us all to the brink of sanity. I had reached that point, inevitable for even the most patient and compassionate of Denali guides (which I was not), when even nine-to-five work seemed inviting. Steve was reconsidering a return to being a roughneck, where he had worked long hours at no small risk to his health on the barren oilfields of Wyoming.

Clearly, burnout is a disease of guides rather than of clients. When push comes to shove on Denali, the conscientious guide, like the captain aboard a sinking ship, will overlook his own needs in order to attend to those of his charges. Consequently, guides become thin, dehydrated or frostbitten while they're haranguing their clients to eat, drink or dress properly. Or they become exhausted because they can't trust anyone else to cook meals, break trail or grope through a whiteout. And if a rescue breaks somewhere on the mountain, it's the guide who evacuates a stranger all through the night, then turns around to tow his clients another 1,000 feet up the mountain.

■ ■ ■

At 17,200 feet, on the eve of our departure for the summit, I told Duane to please shut up. After pulling his ice-cold and rancid bare feet off my stomach, I swore an oath not to ruin my vision of Denali by guiding it ever again.

We left at 6:00 a.m. The prescience of great disaster seemed to hover lower than the lenticular clouds over nearby Mount Foraker. Despite Duane's presence, I prayed for mercy.

Every trip has its saviors. Jack, Fran, Bruce and Walt had put up with Duane for 17 days without complaint. I hoped Duane would become too weak to make it to the summit; then Steve or I would take him back down so his long-suffering teammates could reach the summit in peace. But at 19,000 feet, Duane despite his slug-like pace, could not be persuaded to turn back—it was Walt who felt too sick to continue. Steve volunteered to pull Duane and the others to the summit.

Before I turned around to escort Walt down, I pulled the emergency dexedrine out of my first-aid kit and handed it to Steve.

"What's this?" he asked.

"Speed," I replied, "in case Duane runs out of gas."

Three interminable hours later, Steve finished yanking Duane the final 10 yards to the summit, blessing him with the usual round of heated verbs and graphic nouns. Duane collapsed onto the snow without his traditional reply. Steve reveled in the exuberance of Jack, Bruce and Fran, and they all forgot about Duane as the continent fell away beneath their cramponed feet. As Duane lay unconscious, they had much to celebrate, not the least of which being that Jack had recently turned 63.

When it came time to leave, Duane couldn't move. Steve was shocked. He asked Duane to stand up, but Steve's desperate and gentle prodding bounced off deaf ears. Steve pulled out the dexedrine with shaking fingers, pried open Duane's lips and helped him wash down the pill with a slug of icy water.

Then Steve lit into him anew, unleashing a litany of original verbiage more brackish than anything he had ever heard on the Wyoming oilfields. As if responding to a dream, Duane rose to this fresh layer of indecency and forced himself to his feet, morally outraged, croaking, "Y'all can't talk to a human bean' this way."

Steve figured that if he could just keep Duane moving, the dexedrine might jolt his adrenal glands into action. During the first hour, Duane toppled a dozen times. Steve let him rest briefly, pumped him to his feet with more verbal heresy, and wondered if Duane would force them to bivouac in 30-below-zero temperatures, which could doom them all. The other three lurched down in distressed funks, alternately turning to yell such provocative suggestions as, "You're a loser, Duane!" toward their martyr.

A half-mile below the summit, at the 19,650-foot Archdeacon's Tower, Duane started reacting to the drug and the blasphemic insults. He became a veritable jukebox of complaints, which offset some of the barrage from his companions and turned his face blue. They arrived back in camp 18 hours after leaving, and I spent the rest of the morning monitoring Duane. Although everyone else snoozed in blissful exhaustion, Duane was wired for sound. He would not shut up.

The next day, Steve had bloodshot murder in his eyes. Since it was my turn to deal with Duane, I insisted that Steve take the day off; if Duane didn't somehow kill himself, Steve would gladly do the job. Guiding novices down the most interesting 3,000 feet of the West Buttress ridge and headwall is never relaxing, but with Duane tottering every step of the way, our passage became one of the most angst-ridden trials since Whymper's ill-fated Matterhorn descent of 1865—when one uncoordinated climber slipped and pulled his ropemates to their deaths.

At 14,300 feet, Steve had napped and spent the day kibitzing with other climbers. Since he seemed relaxed again, I offered him the rope with everyone but our minister; the foursome trotted down to the 11,000-foot camp while I chaperoned Duane. Five hundred feet down, I strapped Duane's pack onto my own. At Windy Corner, Duane stepped into the same crevasse he'd stepped into a week earlier.

Duane asked for frequent rest stops. After he caught his breath he issued that patronizing smile and boasted of the glory of his Chief Guide, who had lifted him onto and back down from the summit. In that syrupy drawl Duane schemed of the magnificence of his homecoming. Wiping the drool from his lip, he predicted how all the troubled teenagers in his parish would come to him for advice, "Ah'll say that Ah too have experuminted with drugs." I pondered violating the First Commandment.

I had long concluded that getting to the top of Denali was not worth an iota of blackened tissue, let alone giving up one's life. Duane was probably similar to hundreds of other clients who had been dragged to our highest piece of geography. His physical unpreparedness and naiveté about his own survival, let alone the safety of his companions, had spoiled the trip for most of the clients—whom I now respected immensely for their efforts.

Duane's experience seemed an off-kilter joke to the memory of a safe climber like Chris Kerrebrock, who suffered a cruel, slow and unjust death. If the god of Duane's prayers did exist, Chris would not have been made to wait for three hours to die inside of an open ice coffin, looking up at three vertical miles of cocked avalanche slopes.

When Duane demanded his 50th rest stop of the day, I took action. "To the basecamp," became my credo. In front of 13 stunned clients of Rainier Mountaineering, and 40 yards from Steve's tent, I began dragging Duane with the rope, belly down in the snow. Most of the camp watched in awed silence; Fran and Bruce and Jack and Walt applauded vigorously.

The only voice that could be heard under the otherwise placid sunset clouds was a nasally distinctive, but somewhat snow-muffled, Georgian-twanged judgment: "Jonny, Jonny," the voice whimpered, "Y'all can't treat a human bean' this way!" When I reached the tents, bathed in peach alpenglow, I untied Duane as if unhooking an inedible bottom fish.

Steve passed me a cup of steaming tea, a smoking cigar, and a smile pregnant with congratulations; he would chaperone another dozen groups up Denali. I never guided the mountain again.

prophet or heretic?

Eric Perlman
July 1995

Ray Jardine could be said to have invented modern rock climbing in the U.S.— he pioneered the technique of hangdogging, or repeatedly falling and hanging on the rope to work out extremely difficult moves. He invented the first camming device, the Friend, whose security and ease of placement revolutionized crack climbing. And he established arguably the first 5.13 in the U.S. Like many pioneers, he was mocked for his methods. And he's still challenging our thinking. Imagine, he asks, a bolt-on, plastic-hold, 5.10 route up El Cap.

When Ray Jardine left Yosemite Valley in 1981 at age 36, he felt his climbing career was finished. He believed his days of pushing standards were over, and if he could no longer break barriers he was just not interested.

Jardine had set himself up as a hard act to follow. He discovered and redpointed the world's first 5.13—the Phoenix—in 1977, six years before sticky rubber hit the market. He used his engineering skills as a space-flight mechanics systems analyst to design and produce Friends. And, in his most unsung innovation, he developed the process of "working on a route," a first-ascent method that angered many climbers of his generation, in part because it allowed a large-boned egghead to surpass the athletic accomplishments of whip-thin climbers 15 years his junior.

Jardine angered, alienated or amazed everyone who climbed with him. His methodical whittling down of a route, move by move, tried belayers' patience. A total unconcern for the ethical standards of others earned him enemies. Yet like all powerful thinkers, the man could not be dismissed. The brilliance of his routes, the undeniable contributions of his designs and his yet-unrealized visions of the future of the sport place Ray Jardine among the rarest of climbing revolutionaries.

You put up Crimson Cringe (5.12) in 1976 and the Phoenix the following year. You also pioneered more natural-pro 5.12s in Yosemite than anyone else. Did you ever think you might have been the best free-climber of your era?

No way. Guys like John Bachar, Ron Kauk and Dale Bard could climb circles around me. They had natural ability, but I had a couple of other advantages. Friends, for example—and they weren't a matter of luck. I designed and built them to help my climbing. In part, my new routes were a spin-off of that technology. Also, the cutting-edge climbers of that time had energy and drive, but I might have been more focused. When I set my mind to a problem, I wouldn't let go until it was in the bag. I succeeded not because of natural abilities, but because of motivation and determination and love for what I was doing.

Tell me about the Cringe.

Originally, the crack's lower third was packed with dirt and had bushes growing out of it. Higher, the wall was covered in lichen. After I had thoroughly cleaned it, Kauk and I gave it a try; we managed only the first 10 feet. We went back a few days later and made it another two feet. On the way to the car, Kauk mumbled something about 5.12. He must have figured it was a lost cause. But I wasn't dissuaded, and I went back eight more times and finally succeeded. What became known as "hangdogging" should really be called "Crimson Cringe-ing," because it was there that I first practiced that technique.

What did you think of Bachar's ascent a few days later?

I watched from a distance and could hardly believe it. Bachar literally cruised it, and Kauk followed just as nicely. I felt like I had broken the trail through deep snow, psychologically, but I was really impressed with John and Ron, particularly because they hadn't used Friends. In terms of breakthroughs, their effort was equally significant.

What about the Phoenix?

When I first laid eyes on the Phoenix, I knew it represented a whole new level. It was absolutely stunning. We started just below the traverse, and when we had finally completed the upper section, we pulled the belay station, repositioned it lower and started over. That's why we called it "the Phoenix"—it rose from the ashes a new and much tougher critter. We knew it was 5.13, but rated it 12 because we didn't want to be presumptuous. We thought it might be down-rated later; we let them up-rate it instead.

What was your strategy for ascents?

I would climb as high as I could, hang on the rope and shake out, then try to continue higher, freeing each move and not pulling on pro. I called it "working the route," and I guess the name stuck. I considered it a form of rock gymnastics. My goal for each day was to start from ground zero and reduce the number of resting points. Finally the day would come when I could climb the whole enchilada in one go.

The longer I worked on a route, the more meaningful the first ascent seemed and the more I felt I was pushing my personal frontiers—both physical and mental. So it was far more than just ticking off routes. It was a quest.

In the '70s and early '80s, when the leader fell, he was expected to ask to be lowered to the ground. Working out the moves while hanging in place was considered unethical. Why did you depart from those rules?

I didn't care for the [then-accepted] "trad" yo-yo approach. One person would climb as high as possible, fire in a piece and lower off. Then a fresh climber would tie into the sharp end and toprope to the high point before trying to climb higher. This sometimes went on for days, during which time the pro and the rope were left in place. As far as I was concerned, it was a quick and dirty method of putting a man on top. Had I yo-yo'd my routes, I could have knocked them off a lot faster. For me, my way provided the greatest challenge and rewards.

And you got flack for that?

Of course. It's an unwritten law that free-thinking individuals may as well paint targets on their T-shirts. People are bound to take pot-shots from the peanut gallery. It's healthy to scrutinize and analyze other people's choices, but I think it's more important to follow one's own vision.

Speaking of vision, in terms of purity of ascent, Bachar's style was exemplary. I never saw him yo-yo or finagle a climb in any way. Unlike the rest of us, he was willing to forego the first-ascent arena, and, in this respect, he showed a great deal of imagination, even though it was aimed in a different direction. We were astronauts; he was Siddhartha.

You were a strong believer in training for routes.

We all were. Most people trained on boulders, but I used to go to a bridge on Highway 120. One concrete support had a pair of cracks—left- and right-facing corners—that were perfect for liebacking. I used to clip water jugs to my gear sling and repeat liebacks up and down without touching the ground. I developed a theory of using what I called "primary" muscles for climbing and "secondary" muscles for resting, mostly in a lieback position. When my primary climbing muscles got tired during a tough fingercrack or off-size handcrack lead, I'd kick into a lieback and hang on with my secondary muscles. This would allow a measure of rest at virtually any point. The Phoenix was a good example of this plan in action. There's a great left-leaning pause just below the traverse.

How long did you live in Yosemite?

Eleven years. When I wasn't climbing, I went hiking, mainly to scout new routes. I could tell a climb's rating almost at a glance, and I had pages of potential routes in my notebook, prioritized as to difficulty. I focused only on the more challenging entries, and I wasn't afraid to spend a month working out the moves. Such a thing was incomprehensible to most climbers back then.

You loved the sport, yet you quit in 1981. Why?

Climbing was important to me, but it wasn't my whole life. I wasn't trying to build self-esteem or make something of myself through climbing. I had worked as a space-flight mechanics systems analyst. I was, and still am, a Christian. What attracted me to climbing was the physical and mental challenges, the beauties of the natural world, the companionship of kindred spirits and the freedom to live in a realm far beyond the norm. Global sailing, ultra-long-distance hiking and sea kayaking do the same things for me. I didn't burn out from climbing; I just needed to expand my horizons by pursuing other, equally challenging adventures.

Do the 5.14 sport routes in Europe, Smith Rock or Cave Rock give you the urge to go climbing again?

My palms get sweaty just looking at the pictures of these routes! I go to the local crags and climb 5.10 and 5.11 every few years, but climbing 5.12 and beyond takes a lot of dedication. I still have the dedication, but I apply it to other areas. There's only so much you can do with full intensity. I like to expand my vistas, not just in one activity, but in several. Every time I try something new, I have to start all over, and that's what growth is about.

You were the first to try to free the Nose of El Capitan. What do you think of Lynn Hill's ascent?

Fantastic. She's done a lot for climbing, and for feminism as well, no doubt. I think she deserves all the credit in the world.

What's not so heartening is that although feats like Lynn's are impressive and motivational, they exclude the average climber. Even the Phoenix, done 17 years ago, still excludes the vast majority of climbers. I don't see climbing as an elitist or spectator sport. I would hate to see it wind up like football, where everyone sits around watching the pros. The bottom line is that establishing hard routes is not taking us into the future. Making routes of moderate difficulty in accessible and spectacular places will.

When somebody makes a free route up the South Buttress [on the Nose] of El Cap at 5.10 or 5.11, for example, that will be a monumental achievement. There are big chunks of rock throughout the world that offer virtually unlimited potential. Why not get creative and open them up for lots of climbers to enjoy?

Are you advocating an aggressively manufactured route up the Nose?

The Nose had already been aggressively manufactured prior to Lynn's free ascent. The first four pitches, for example, are climbed mainly on pitoned-out pockets, chiseled unintentionally by aid climbers. Lynn surely clipped a lot of bolts along the way. What about my traverse into the Grape Race? She used that, didn't she? What I'm advocating is opening our eyes to reality. Climbers often surprise me with their ignorance of the amount of sculpting that has gone into a great many classic routes.

Back then, it was no secret that the more times a pitch was aided, the more pitons were hammered in and out of a thin crack, the wider and more usable the pin holes became. Eventually, the holes would accept the fingers, and voilà! The pitch was opened for free climbing. Aid climbers didn't blame themselves for modifying the rock.

In the same vein, Jim Bridwell didn't blame himself for accelerating the process. His technique, known as "pinning out a crack," provided some of the finest routes Yosemite has to offer. We joked about the "thank Bridwell" holds, but we certainly enjoyed climbing the routes they allowed. In that sense, Bridwell was a visionary, for in manufacturing those routes, however slightly, he opened them to free climbers. And in so doing, he enhanced Yosemite's climbing scene.

Of course, not all of Bridwell's routes were tooled. Not by any means. Nor were mine. But I think we both had the vision to know when something needed to be done, and we both had the courage to do it.

So what happened on your attempt to free the Nose?

I worked on it for four months in 1981, freeing all the lower sections and making no modifications until the traverse near Dolt Tower. I first spotted the traverse from the ground, using a high-powered telescope. When I finally reached that point, I found a line of shallow, sloping pockets leading left to another crack system: the Grape Race. It looked like 5.11, but after several days of working the traverse, I determined that it was a lot harder. So I bought a cold chisel and made it 5.11.

Tooling those holds was an experiment. I had never modified face holds, and I didn't realize how blatantly they would stand out. Like everyone else, I was pretty appalled by the results. In retrospect, I should have experimented elsewhere and made the holds smaller. Still, there's the question: Was I committing a moral injustice or making a little bit of history?

Anyway, my vision was for a moderate free route up the South Buttress of El Cap. I wanted to make it not of the highest standard but of the highest meaning. After I realized that enhancing face holds was not the way to go, I quit the project—I knew its time had not yet come. Progress demanded that the route go free, but I lacked the technology to contrive the necessary holds.

It's gone free now.

Yes, and like everyone else, I'm very impressed. But I'll be even more impressed when the wall is opened for the masses.

Are you suggesting that Lynn's route be modified to accommodate climbers of lesser standards?

No. I'm talking about a route I call Numero Uno. From the beginning, this wasn't meant to follow the Nose exactly. The traverse into the Grape Race is one example. I knew that the Great Roof pitch would be extremely difficult, and, because it's usually wet, it wouldn't make the best line. So to bypass that, I was looking at the system to the right, starting from the Gray Bands just below Camp 4 [the ledge, not the campground].

Even though I had to give the route up, the four months I spent on Numero Uno were among the highlights of my climbing career. I absolutely loved it up there.

A manufactured route for the masses would be more than a little controversial today. How do you feel about that?

I think many climbers are living a double standard. They consider themselves purists, condemning some forms of rock modification and condoning others. They pause on the ascent and moralize about how the rock should be left in its natural state, then reach up and clip into a ½-inch bolt drilled three inches deep and proceed to climb a sterile 5.11 crack that was formerly packed with soil, nurturing grasses and the occasional clump of cheery flowers. Along the way, they smear this denuded crack with magnesium carbonate. And here's the funny part: The person who prepares the route catches the flack, while the person who climbs it catches the glory.

I'm not saying that we should give everyone *carte blanche* to start modifying rock. Quite the contrary. We have to make a distinction between the visionary and the hacker. One is characterized by perception and motivated by love for the sacred stone; the other is motivated merely by ego.

A lot of people aren't going to want to hear this kind of talk.

Yes, it's difficult to see things change, particularly for those who have a lot of time and effort invested in the old ways. This happened to me with the transition from EBs to sticky rubber, when suddenly all my routes were made a little easier.

I'm not trying to stir up a hornet's nest, but if the idea of openly manufacturing routes seems bizarre, consider this: The standards we used in the '70s to establish new crack routes were consistent with the short supply of difficult cracks. Face climbers limited themselves to placing bolts by hand on lead. This was an ethical standard designed to enhance the challenge and adventure.

Eventually, though, most accessible routes that could be climbed in this manner were conquered. Progress demanded a change of standards. But woe to the first people with the vision to start placing bolts on rappel! They were all but tarred and feathered. In retrospect, we see that they were right. Imagine what places such as Smith Rock would be like if we hadn't evolved beyond those old rules.

If placing bolts on rappel opened up vast possibilities for climbers, then applying artificial holds will probably do the same, on a far grander scale. Let's face it. There are only so many natural lines. The top climbers are pushing the standards, but they are not expanding the breadth of possibilities for the average climber. We have a great many enormous walls, such as those found in Yosemite, with a very limited number of routes on them. Most are aid routes—and rather dangerous ones at that. I think it's inevitable that people are going to start gluing on holds and making free routes. As it stands now, El Cap is basically the domain of the aid climber. This is regrettable, as free climbers are in the vast majority.

I think a major glue-job on El Cap would trigger an outcry . . . or worse.

Outcries are precursors to change. When the carabiner was introduced, the majority of climbers were outraged and labeled the free-thinkers who used them cheaters and scoundrels. Ditto with nylon ropes, Friends, hangdogging and placing bolts on rappel. Changes of this magnitude have always left the community struggling for equilibrium. But so far the sport has always pulled through and moved ahead.

A line of artificial holds up 3,000 feet of virgin rock might upset people, but suppose there were just a few sections of 10 or 20 holds that linked multiple pitches of natural climbing—something like a free alternative to the bolt ladder between the top of Texas Flake and Boot Flake. In the politics of today's climbing, that might represent an acceptable concept.

Where does this path of route fabrication take us?

Into the future. Of course, there are lots of social, legal, ecological and ethical considerations that are going to have to be dealt with before a route like Numero Uno will exist. But I think it's inevitable. With technological advances, holds won't be chopped or drilled into rock—they'll be glued to it, probably with thermoplastic adhesive. In all likelihood, holds would be little blobs of compound, colored and textured to blend in. And here's the most important part: They would have to be removable without defacing the rock. Any inventor-climbers out there looking for a golden opportunity?

After freeing half the Nose, you quit climbing. What have you been doing since then?

I left Yosemite in 1981 a controversial and wealthy figure. The wealth I'm referring to was the many years of gloriously rewarding climbing. In terms of having lived life to the fullest, I consider myself a bloated plutocrat. I also had some spare change in my pockets for the first time, as Friends began to find acceptance.

I moved to San Diego, rented an office and started marketing Friends to climbing shops. My main objective, though, was to pursue my lifelong dream to sail around the world. After many months of searching, I stumbled upon one of those once-in-a-lifetime deals on a 41-foot blue-water ketch. With Jenny, my partner of 14 years, I sailed off for the fabled Marquesas Islands, thousands of miles beyond the horizon.

Were you an experienced sailor?

No. We had never even reefed a sail. But I had studied ocean voyaging and sailboat handling, and I was confident I'd chosen a well-found vessel. And we had spent a lot of time beefing up its components. We had stowed a year's supply of food and a full complement of spare parts, and beyond that, we figured we would just learn as we go.

Eleven days into the journey, we ran smack into our first hurricane. It tossed the boat around like a cork in a washing machine, leaving us hanging on below deck in sheer terror. The fury of such a storm is unimaginable, and, unlike with climbing, you can't rappel off. For three days, we existed in the netherworld. Then the storm passed, and we went on to complete the 30-day passage to the first island—the first leg of a three-year trip across the Pacific, Indian and Atlantic oceans.

What was the highlight of the journey?

Five months in South Africa.

Why?

The country is gorgeous, the wildlife is astounding, and the people are very friendly. Then we sailed around the Cape of Good Hope, also called the Cape of Storms. It was 5.13 sailing, but we made it around safely.

Do you still have the boat?

No, by the end we were a little sea-weary. We sold the boat with the intent of hiking the Pacific Crest Trail. It was too late in the season to start, so we spent the summer hang-gliding in Utah. I flew mainly cross-country, talking with Jenny by radio and directing her where to pick me up.

Sailing hadn't done much for our leg strength, so we put in a long winter of training on Pikes Peak. By spring, we were ready. Standing there at the Mexican border with 2,600 miles of trail ahead of us was like standing at the base of an extremely large and difficult rock climb. We had both done our share of backpacking and had prepared ourselves as much as possible. But neither of us was sure we could hike that far in one go.

As expected, it didn't prove easy, hiking 15 hours a day with heavy loads, but it wasn't work either. Each evening we crawled into the tent pleasantly pooped and ready for another day. And after the first three weeks or so, we began to adapt, and the journey became a real joy. The more time we spent out there, the more we tuned into life. Climbing had always been an up-and-down process. But long-distance hiking is always up. We didn't just experience the wilderness for a few hours or days at a time; we lived it 24 hours a day, all summer long. And what surprised us was how much wilderness there is between Canada and Mexico.

Two thousand, six hundred miles wasn't enough hiking for you?

It just whetted our appetites for more. We've now done the Pacific Crest Trail three times and written a book on the subject, *The PCT Hiker's Handbook* [AdventureLore Press, LaPine, Oregon]. In 1992, we hiked the 2,800-mile Continental Divide Trail from Canada to Mexico via Montana, Idaho, Wyoming, Colorado and New Mexico. I wrote a small book on that, *The CDT Pocket Planner*. The next summer, we hiked the 2,100-mile Appalachian Trail. Last year, we founded the American Long-distance Hiker's Association and started publishing a newsletter.

What attracts you to long-distance hiking?

We love spending the summers in the wilds, and we especially enjoy the physical challenges of hiking marathon mileages every day over rugged terrain. The fact that you have no car waiting back at the trailhead sets your mind free. You can just keep going, day after day, for months on end. Genetically, I think people are programmed to live and survive in the wilderness. Mega-trails elevate life to this level. And the remarkable part is that they exist virtually in our own backyards. We don't have to travel halfway around the world to hike them.

On your last hike of the Pacific Crest Trail, you and Jenny covered more than 30 miles a day, day in and day out. You just turned 50. What's your secret?

Minimize your gear weight, train gradually, and stay positive. A close friend, Brice Hammack, turned 74 last summer while completing the triple crown of long-distance hiking: the PCT, CDT and AT. He's my exemplar when it comes to keeping active.

How long does it take you to do these trails?

Most through-hikers take five or six months. Last summer, Jenny and I hiked the PCT in three months, four days. We go with the minimalist approach and custom-make our packs and gear in order to lighten and simplify our loads. Our packs weighed 8½ pounds each, fully equipped.

What!?

Not including food and water.

How much food and water on average?

We carry about 2½ pounds of food per person per day and resupply on the average of every four or five days [140 to 150 miles]. We try not to carry water, except in the deserts. Normally, we filter and drink creek and spring water.

When did you take up sea kayaking?

I was introduced to it back in the '70s and did seven big trips in the Sea of Cortez, along the inside coast of Baja. After Jenny and I hiked the Pacific Crest Trail the first time, we wanted to return to the seafaring life. But this time we didn't want to isolate ourselves from the land as much as we had in the sailboat. So we bought a two-person kayak and trained for paddling all winter. In spring '88, we launched the boat at Anacortes, Washington, near the Canadian border, and paddled the Inside Passage 1,000 miles to Skagway, Alaska. There we disassembled the boat and carried it and all our gear over the historic Chilkoot Trail, across the Divide and down to the first ice-free lake. That was about 100 miles of hiking in three portages. When we reached the headwaters of the Yukon River, we reassembled the kayak and paddled another 2,000 miles to the Bering Sea. The trip took 100 days.

You're planning to paddle north of the Bering Sea this summer. What are your objectives?

When we reached the Bering in '88, we felt a strong urge to keep going. God willing, this next trip will be a continuation, leaving no gap unpaddled. We're hoping to do about 1,500 miles from the Yukon River Delta, around the top end at Point Barrow at 71 degrees north, and as far eastward along the north shore as weather and pack ice permit. The trip should be a blend of extremely cold water, down to 34 degrees; barren coastlines; harsh weather and frequently heavy seas; stretches of pack ice; midnight sun; lots of sea life and wildlife, including grizzly and polar bears; and always a more distant horizon. In short, another fine adventure.

off the wall

credibility gap

Greg Child
March 1986

No active writer on technical climbing is as accomplished as Greg Child. His stories and books on pioneering climbs from Yosemite to Pakistan are rife with mordant humor and flashes of insight. In this fictional story, Child writes of expeditions and relationships gone awry—and about one very bad assumption.

That old coot, bush pilot of the Alaskan wilds, christened Fabian Cootes by his mother, but known among the drinking class as simply "the Coot," pointed to a break in the clouds blanketing the Kitchatna Spires, sprang his thumb upright, then eased forward on the controls of his Cessna. The airplane dropped into a misty tunnel, and the Coot's three passengers felt their stomachs rise as if on a rollercoaster.

The Coot's hedgehog-stubble head darted from side to side with wide-angle eyes that looked to avoid precipitous valley walls, and he flicked the altimeter and its goddam nasty habit of sticking. A sunbeam poured through the cloudhole, overtook the Cessna and bounced off the glacier. The Coot cackled and yelled, above engine-drone and rush of wind-over-airfoil, that they were landing.

They had been circling over the Kitchatnas for what had seemed hours to the nervous landlubbing passengers, looking for that break in the clouds that would allow the plane to descend like a weevil through a hole in Swiss cheese. The problem with the Kitchatnas is that the range is so far from Talkeetna airstrip that a fully loaded plane can only make a few exploratory passes before finding the right glacier to land on. The Coot had kept shaking his head at the cloudbank below, muttering "I dunno" to the fuel gauge. He had also kept asking Follett to look at the map and remind him of the elevations of the highest summits, while he held the Cessna two hundred feet above that, interchanging glances between altimeter and window. Follett felt like a sailor scudding over murky shallows, watching hungry-as-shark reefs waiting to tickle the hull of their ship, while the senile Captain ordered depth soundings from the charts.

Suddenly they were cruising beneath the cloudmass, a few dozen feet above the snow, flanked by cloud-capped walls. The valley ended in a box canyon, blocked by the misty ramparts of a black-footed mountain. The Coot studied the lay of the glacier for a place to put the plane down, but the light was flat, devoid of relief or definition. The opening that let them enter the valley closed tightly behind them, swallowing them like a morsel of food.

From beneath his seat, the Coot grabbed a plastic garbage bag of clipped spruce boughs, each about two feet long. He shoved them into Follett's hands.

"Gonna make a tight turn at the enda this canyon, pass over this stretch beneath us, turn again, then put 'er down. While we make that first pass you open the door and drop a spruce bough, one every second, till they're all gone. That'll gimme somethin' to go for. Got me?"

"Gotcha!"

As the black-footed mountain reared up, the Coot banked the plane sharply. Follett looked bug-eyed at the proximity of the rock walls, noted water streaks, cracks and the arching architecture of granite, even detected a drab dab of moss clinging here and there.

"Incredible!" he shouted.

"I'd rather watch it on T.V.," said Bernice sickly. From the back a staunch smell of puke seeped forward. The sensations of gravity were more than she could handle, and she was losing her lunch into the nearest receptacle, which was her husband's (Wilson's) climbing helmet.

"Crack open the door 'n start droppin' them boughs!"

Throttled-back engine noise filled the cabin as Follett let the greenery fall until the garbage bag was empty and then that too went out the door. Thumbs up from the Coot, a less dramatic turn at the more open, yet heavily crevassed lower glacier, then a slow glide toward the spruce-bough landing strip with the engine gurgling like bubbling soup.

"Hold onto that puke-bowl girlie, we're goin' in!"

"Stop calling me girlie!"

It was impossible to tell where ground began and where it ended. Light was so diffuse that air melted into snow. The spruce boughs hovered as if in space, though in fact, they sat firmly on snow. The Coot aligned the plane with the boughs and eased it down. They awaited the soft wumph-bump-rattle of skids hitting snow. Instead, *terra firma* came up like a slap in the face, with a neck-wrenching jerk, whine of engines suddenly gunned and a deluge of powder snow spraying over the windows.

"Christ! Watch that puke!" yelled the Coot.

They had hit deep powder and before they became bogged in it the Coot opened the throttle to pull them out. The Cessna emerged onto firmer snow, still in the line with the spruce boughs. The Coot gave another mad cackle. Wilson felt the talons of Bernice's fingers grip his kneecap painfully.

They settled into a rollicking ride over flat glacier, breathed easy, slowed, and just as the Coot was about to open the throttle one last time to charge the carburetors, they crashed.

A narrow crevasse opened beneath the tips of the skids, just enough to make the plane lurch forward. Again, the Coot gunned the throttle and jerked back on the controls, to bounce the plane out of trouble, but a wingtip dug into the snow, spinning the plane around and causing it to come to rest, nose down, tail up, tilted sideways.

The Coot let out a mouthful of curses and dived out the door. Puke filled the cabin like a coat of sour paint, soon to be replaced by the volatile smell of aviation spirit, leaking from the wing-mounted tank. Cold, cold air greeted them as they crawled over displaced baggage and onto the snow.

"Welcome to the Kitchatna Spires," said the Coot sarcastically, holding his ridiculous fur hat over the fuel cap to stop the spillage, while pearls of snot froze in his nostrils.

Follett tucked his hands beneath his armpits and looked about him. He had crashed on a glacier in the center of the Kitchatna Spires with Wilson, Wilson's wife and a mad bush pilot. Bernice's face was a portrait of dismay. She made a few steps into the soft snow, sank to her thighs, backtracked, toppled over and emerged completely white. Her eyes welled. Wilson put his arm about her. She slapped it away.

"I knew it would be like this!" she whined.

Follett thought again, more concisely. He had crashed on a glacier in the Kitchatna Spires with the above personnel, a storm was brewing, they would be here for three weeks to try to climb a mountain they couldn't even see because of the clouds, and Bernice and Wilson were breaking up, had been breaking up for months, holding out against the dread realization that they couldn't stand the sight of one another, trying to salvage their relationship with crackpot schemes like Wilson's idea of bringing her to the icy wilds of Alaska where they could be together and away from the factors they blamed for their disenchantment with one another. They had made a big mistake in letting it get this far. The Coot had made a colossal mistake.

"Let's get this muthah unloaded and back on her feet or she'll spill all her fuel and I'll never get outa here," said the Coot, shoveling out the wingtip.

"I wish you'd stop referring to that piece of junk as 'she.' It's inanimate. Why don't you call it an it, or a he?"

The Coot stared at Bernice to be sure she was serious.

"Cyclones, mountains, ships, planes and cars all been 'she' ever since I could crawl. Can't grow balls on 'em overnight."

Bernice turned away. The Coot winked at Follett. "Good luck, son," he said. "You'll need it."

■ ■ ■

An hour later the plane was upright and the Coot made haste into a brief hole in the clouds.

They pitched two tents in falling snow.

"Hey. How do you know we are on the right glacier?" asked Bernice through chattering teeth.

Wilson and Follett pondered this. All they knew about the place was the photograph they had seen of their mountain in a magazine. They had not seen any landmarks well enough to be certain that this was the right glacier at all, but had taken the Coot's word for it.

"Coot knows this place like the back of his hand," said Wilson authoritatively.

"How do you know?" tested Bernice.

"He said so."

"You believe that old drunk?"

Wilson looked around him, smiled heartily and said most sarcastically, "Got a better idea, babe?"

As they entered their tents and zipped them shut Follett heard the icy cautioning of Bernice to Wilson, to stop calling her babe.

■ ■ ■

What began as a questionable idea soon turned into a bad one, as five days of storm set in. They had been warned about Kitchatna weather. Two feet of snow fell in one day. And there was the problem of daylight; there was too much of it. The Arctic day was long, and soon all notion of time, as punctuated by the division of light and dark, was lost. Remaining true to their inbuilt biological clocks, they managed to sleep away the brief night, so they never saw blackness. For Follett, alone in his tent, it was one long day. For Wilson and Bernice, it was one long argument.

Wilson and Bernice had been married for two years. She was inclined toward a feminist individualism that had always seemed an unlikely alliance with such a dyed-in-the-wool chauvinist as Wilson. His justification for the mismatch was that opposites attract, hers that given enough time the infidel Wilson could be retrained. Follett had never fathomed the emotional mechanics of the quarrelsome pair, and initially gave them no more than a month before they either killed each other or filed for divorce, but there they were, two years into the bargain, still arguing over those trivialities that are inexplicable to an outsider, but of overwhelming importance to a couple.

Bernice had never lived on the snow before. She had a low tolerance for camping. She hated dirt, wild animals and body odor. Climbing was a waste of time and mountains a waste of space. Her world was the world of the city: plays, art galleries, concerts, good restaurants. Freeze-dried food had to be another of Wilson's elaborate pranks, but alas, it was all too real, and there were weeks of it.

What then, was she doing in a place so alien to her nature? If this was a last-ditch attempt to revive a dying relationship, it was like giving an anemic a transfusion of Tabasco. Follett had learned that Wilson's mind worked in convoluted ways, and that he performed best under stress.

The logic was that they needed to look at their relationship objectively, in a place removed from the trappings of society. Wilson blamed their failures on the influence of her wrong-headed, feminist friends, who whispered in her ear that Wilson, as a male and provocateur of all that was wrong with the world, was best dispensed with. On the other hand, Bernice held that Wilson challenged her every idea as a matter of principle, a trait inherited from his redneck, farm-belt father.

They could debate this for only so long before histrionics and shouting took over. It was a match made on the far side of heaven.

■ ■ ■

When the clouds cleared and the mountains were revealed they found that the Coot had, after all, landed them in the right place. Their peak stood beside the black-footed mountain, separated by a saddle known on the map as Credibility Gap. The icy couloirs of their route shone like glass. The two climbers made ready to leave.

Avalanches sprang from the sides of the valley, encouraged by the sun to cut loose from their moorings. As they packed, they watched plumes of snow cascade down the walls and across the glacier.

"You'll be alright Bernice—it's safe here. Just don't go too far from the camp. There are crevasses behind and avalanches ahead. Think of everything beyond a fifty-foot radius of camp as a minefield."

"You mean I'm trapped here while you go off on your climb? What if you don't come back?"

Wilson sighed. "There's plenty of food. Just wait for the Coot. We'll be alright. Just do the housework while we're gone."

Those were the sort of remarks Wilson couldn't help, as if his mouth had a cantankerous will of its own, or an unseen ventriloquist inserted the words to cause trouble. He immediately regretted his remark and shrank against the coming vitriol.

"Would you like me to make you a nice sandwich for lunch, hmmm? Wash your underwear while you're gone? Brave men, off on your argosy, leaving frail woman in the nest to do the dishes. Listen, I'm not your mother." Her tone was acid enough to melt the snow at her feet.

They saddled their packs. Bernice proffered her cheek for Wilson to peck, like a wife bidding hubby good-bye as he went off to work to earn the daily bread. Their body language was a theater of sarcasm.

A hundred yards from the tent, Follett turned to Wilson.

"You enjoy it, don't you? Both of you like to draw blood. Most people come to places like this for the peace and quiet, but you, you import the conflict, bring it with you, let it ferment, fester, and become a plague."

"Relationships, Follett, are a life's work," said Wilson sagaciously. They talked against the squeak and slide of bindings and skis on cold snow, while the mountain grew closer.

After several hours they stood on Credibility Gap. The weather was holding. They entered a narrow couloir, swinging ice tools and cramponed boots into the plastic-like surface until their toes throbbed, then became numb.

But the sky soon darkened and wind began to whip. The storm had merely ebbed like a brief tide. Soon they were retreating down the mountain, being blown around on the ropes by a gale-force wind. Back on Credibility Gap they paused to rest.

"Better get down to basecamp," said Wilson. "Bernice will be worried." Follett wasn't sure if this was sarcasm or simply optimism on the part of his partner.

■ ■ ■

A titled lady was said to have remarked to Sir Winston Churchill, "If I were your wife, I would put poison in your tea," to which a Churchillian bulldog scowl replied, "If I were your husband, I would drink it." And so it was with Wilson and Bernice, as the second storm raged outside the tents.

Follett was sleeping quietly alone when Bernice burst into his tent, pushing her sleeping bag and foam pad into the entrance before her.

"I'm moving in. I've had it with that creep."

"You're just like your mother," yelled Wilson from across the void.

"And you're a clone of your father!" Bernice screamed back.

"What's wrong now? No—don't tell me. I wouldn't understand." Follett drifted back to sleep as Bernice shuffled around trying to get warm and comfortable. He was dreaming fitfully when he felt Bernice's body wrap snugly around his. His eyes opened. His head turned slowly to verify the fact. He registered the impulse to move away, but lay rigid, like a man with a venomous snake curled beside him.

Bernice detected the jump in his heartbeat. "You're nothing but a source of body heat to me, Follett, so don't get any ideas. Just go back to sleep and keep generating heat."

"Well Bernice, I'm glad I can be of service to you. I hope that if the Coot forgets where we are and we run out of food you'll utilize my body fat and eat me."

"Count on it, Follett, count on it."

■ ■ ■

They shared heat for the duration of the storm. Boredom grew like a virus, until one day Follett awoke with the urge to indulge in the same game that Wilson played. Stir-crazy in their confinement, he began to verbally bait her, playing his own version of Virginia Woolf's game, "Get the Guest."

A student of psychology, he was well armed to assail his lodger, having with him a selection of texts. Within those pages lay the stuff of warfare.

"Did you know that Theodor Reik said that 'A woman wants to be recognized for what she looks like, a man for what he does.'"

"Ancient history—bullshit concocted by men."

Follett smiled, thumbed a few pages further till he found another quote. "Ah . . . Thoughts about housework. 'It is not accidental that rooms are representatives of the female body in dream symbolisms and other productions of the unconscious mind . . . The displacement from her body to the room reveals itself in the meticulous attention women bestow upon their living quarters, which are, so to speak, a part of themselves . . . "

He was cut short.

"This is criminal! Stop it! You're worse than Wilson. I'm shipwrecked with a pair of chauvinists."

"Only trying to make conversation . . . " he said with mock mildness.

"Who is this quack, anyway?"

"One of Freud's star pupils."

"Freud! No wonder. Half mad with penis envy. Like a mountaineer and his mountain."

"Oh, come now, Bernice."

"Men's thoughts on women—pah! Psychology is a conspiracy to enslave women. Women will achieve greatness with or without your sexist textbook platitudes."

"Some great minds would disagree."

"Like who?"

"Schopenhauer said, 'Women . . . '"

Bernice shrieked, unzipped the door and dragged her things into the driving snow. She loped through a deep drift to Wilson's tent.

"Moving back with Wilson?"

There was a brief hubbub and a sound of scuffling. Wilson appeared suddenly at the door of Follett's tent, looking bewildered in his longjohns and with his sleeping bag draped over his shoulder.

"No. He's moving in with you." Bernice called, closing the zipper.

Wilson stumbled into Follett's tent.

"Whatever you said to her it must have been good. She only lasted two days with you."

As the day grew on, the temperature plummeted. Wilson bet Follett a six-pack that Bernice would be back before long, frozen and desperate for body heat. Follett declined to bet, which was just as well, for as the big white moon rose over the jagged spires of the valley rim, she came stomping over, and like Moses parting the Red Sea, forced her way in between her two antagonists. Thus ended the game of musical tents.

■ ■ ■

The second attempt on the mountain ended in another storm. Upon their disgruntled return to camp, after three days away, Bernice was curiously inquisitive as to why they had failed.

"Credibility Gap. Hmmm. How strangely apt." She rolled the words around in her mouth like a good wine. "There does seem to be a credibility gap regarding your ability to climb this mountain. Perhaps you boys ought to change your objective to something easier?"

Follett prickled, became defensive to her schoolmarm tone, feeling as if his competence, his very masculinity, was on trial. They had retreated for good reasons. Anyone would have done the same. She was playing "Get the Guest." In any case, what did she know about climbing? Nothing. She was along for the ride, for, for . . . why was she here? Why had she come to this desolate place? She had said a million times that she hated this glacier, was bored to death by it. He began to comprehend the seafaring taboo that women were bad luck on ships.

Stop! He shook his head. Psycho-sexual politics had gone too far. Role playing, role reversals, ego and pride had gotten the better of them. She was a thorn in their sides and they were a thorn in her's. She was a cerebral and political creation whose world revolved around culture and women's issues. They were physical oxen, explorers, seekers of adventure. Men were hunters, women nest builders.

No, no, no! Reductionism, banal as a cigarette advertisement. Don't categorize. All was lost if they fell into sexual stereotyping. He wasn't a chauvinist. Even Wilson wasn't a chauvinist. They wouldn't have let her come if they were. If there were a rock around he would dash his head against it.

"We gotta get away from this crucible, back onto the mountain, or fly away with the Coot, but let's get away before we go mad," Follett said aloud. They stared at him unaware that the situation had become so overblown in his mind. Bernice handed him a cup of coffee and poured a nip of brandy in it. More than a cup of coffee, this was a peace offering, a signal. He looked up at her and she looked back. Had she read his mind?

In her suddenly soft, feminine eyes he detected a hidden capacity for sympathy. He remembered how his mother used to scold him as a child, then forgive him soon after. A rush of chemistry thawed within him. He looked into her eyes and that smile that could launch a thousand ships, and saw something he shouldn't, couldn't believe. Desire.

He gulped the coffee and looked away. A break in the clouds dappled the snow with sunlight. The glacier boiled and shimmered. A mirage. Sometimes in the mountains there are mirages, and you see things that are not really there.

■ ■ ■

That patch of sunlight took three days to grow into good weather. Under cloudless skies they returned to Credibility Gap. It was the last chance they would have to make their climb before the Coot returned.

The wind had blown the fresh snow off the face of the mountain, leaving the ice and rock in perfect condition for climbing. They gained height rapidly, and after twenty hours of continuous climbing were so pleased by their progress and certainty of success that they paused to rest on a ledge a few hours from the summit.

Follett sliced off a piece of salami and handed it to Wilson who lit the stove. Mountains bristled over mountains and basecamp could be seen as a small central dot with footprints radiating out from it, like the spokes of a wheel.

"Follett, I want to apologize for inflicting my personal life on you. I don't know what I was thinking when I brought Bernice along. I thought it would be good for us. Maybe it has been. Our aggressions are certainly out in the open. What I'm sorry about is that you're caught in the middle, in an emotional no-man's-land."

"No need to explain."

"I just don't want to lose her."

Follett nodded, then caught sight of something below.

"Hey, what do you suppose that dark thing is down there? See it? Wandering up the glacier?"

Wilson squinted. The sunlight was bright against the snow, but he agreed there was an object moving around below, making scanning movements back and forth over the snow.

"Maybe it's an animal."

"Up here?"

"Sometimes critters wander up the glaciers from the woods. I've heard of it before. Elk, moose, bear."

"Bear?"

"Sometimes."

"Grizzlies?"

"Sometimes."

The object continued, tacking like a yacht making way against a headwind, back and forth across the breadth of the glacier, a small object in comparison to the huge mountains around it, but clearly as big as a man, if not bigger. Its movements were intelligent. Each time it reached the steepening slopes of one side of the valley it would turn sharply and traverse to the other side, then turn again.

"What kind of animal moves like that?"

"Damned if I know, but it seems to be looking for something. Hunting."

Wilson looked at Follett, a concerned expression on his face. They were both thinking the same thing. If it was a bear down there, then eventually it would reach basecamp and Bernice. Each of them thought out worst-possible-case scenarios.

Why the animal was behaving in the manner it was, zig-zagging back and forth across the valley, was open to conjecture. Follett suddenly worked it out.

"The thing must be snowblind—it's wandered up from the forest and has gone snowblind."

"And now it's using smell and touch to find its way around. That would explain why it turns about when it reaches the steeper snowfields," added Wilson.

The thought of a snowblind grizzly, its head throbbing with the poison of glare, hungry, confused and scared, sniffing its way into Bernice's tent, produced terrible mental pictures.

"Bernice wouldn't stand a chance . . . " said Wilson.

"She could lash it with her tongue," suggested Follett.

"That's my wife you're talking about!"

Both stared at the summit, bright with sunlight, then back down to the advancing predator.

"I'd never forgive myself if something happened to her. She's so helpless down there. I . . . well . . . I love her. Sorry Follett, but we gotta go down."

Follett was speechless, but in agreement.

They rappelled as fast as they could fix anchors. In two hours they were back on Credibility Gap. The creatures's pace had slowed, probably due to weakness and starvation, Follett hypothesized. It was about a half mile from basecamp and closing. Its tack had narrowed as it picked up the scent of camp and began closing in.

"Start shouting. Maybe we can scare it away. At least let Bernice know, get her out in the open where she can run," said Wilson. As they glissaded the snowy slopes they began to whoop and scream. Bernice emerged from the tent and stood staring at the two sliding down on the seats of their pants.

They reached the skis at the base of the mountain, stepped into the bindings and began pacing out the last of it to camp. The animal turned and headed toward the source of the noise. Wilson and Follett stood their ground, wondering which way to turn.

"Bears can't run uphill. If it charges, ditch your skis and run back up the slope," said Follett.

"No, it's the other way around. They can't run downhill. Better to outpace it. It's blind, remember. Keep upwind so it doesn't smell us."

Surface winds whipped hither and thither. The sun dropped over the edge of the mountains and cast long shadows of flat light across the glacier. The creature disappeared. Follett felt his heart pound and sensed what it was like to be hunted. Their eyes scanned side to side, their breath condensed in clouds that hung suspended before their mouths. Where the hell was it? Over a rise in the snow the figure of Bernice appeared, skiing toward them, all smiles and waving with one hand while she poled with the other.

A gust of wind blew spicules of snow into their faces. They shielded their eyes. When they looked up again, they could see the creature behind Bernice, advancing at a gallop.

"My God!" cried Wilson.

He began to ski at breakneck pace toward Bernice, while Follett unclipped from his bindings and pulled his ice axe from his pack, charging on foot, like a Norseman headed to Valhalla. Both yelled ferociously while Bernice stood openmouthed at the spectacle, unaware of what approached.

The thing bypassed Bernice and headed straight toward Follett, who stood his ground momentarily, then let out a whelp and began bounding away. Wilson hurled his ski-pole like a javelin. It appeared to bounce off the creature, and fell lamely into the snow. He grabbed his ice axe, turned about on his skis and followed the shape into the ground blizzard.

Follett spun around to see the horrible dark shadow of the thing fall across him. And behind it was Wilson, his heroic, loyal friend, ice axe raised above his head. The thing collided with Follett just as Wilson crossed his ski tips and tumbled with a dead man's scream into the bloody maelstrom. His ice axe swung down a bit into something soft. Follett felt the first incisor penetrate his thigh and heard the echo of his own terror run the length of the valley. He tried to remember a prayer before being mauled to death. A tumult of wind, snow and screams filled the moment, then suddenly all was silent and still.

Wilson wiped snow from his sunglasses and looked at Follett, who lay on his back, sweat-soaked and panting. He carefully extracted the tip of his ice axe from Follett's thigh. It had penetrated an inch. Blood oozed from the wound and steamed in the cold air.

Between the two of them, wrapped around Follett's legs, was a plastic garbage bag—the one Coot had filled with spruce boughs, and that Follett had tossed out the door of the Cessna the day they had landed.

Bernice appeared from the swirling flatness, looking agog at the imbroglio before her.

"I knew I came here with a pair of crazy sons of bitches, but you two take the cake. I thought all the screaming was because you'd climbed your stupid mountain, but what do I find but you chasing each other around fighting in the snow. I even brought the brandy to celebrate." She looked at the wound on Follett's leg and handed him the bottle. Murmurings from deep in their gullets strove to explain. Wilson held up the bag.

It had all been for her. They had given up the summit to fend off a grizzly. It had looked like a grizzly. . . . Nothing more than grunts came out.

It was a ridiculous story, best left untold.

"Now you two have a drink and make up. We've still got three days before the Coot picks us up and I'm damned if I'm gonna put up with you two fighting, for whatever reason."

They passed the bottle back and forth, twice each, then Wilson patted his friend on the shoulder.

"Forgive me, buddy?"

"Yeah, I forgive ya."

"Good," said Bernice. "And pick up that trash before you come back. This glacier is looking like a pigsty."

revenge

Jeff Long
November 1991

For a sport so rich in literature, climbing is surprisingly bereft of excellent fiction. Jeff Long is one of the few authors whose stories have won respect beyond the climbing world. His novels Angels of Light *and* The Ascent *are serious works that won't set a real climber's teeth on edge with a melodramatic approach to the sport. In this story, set in Yosemite Valley, Long shows us that recapturing the passionate emotions of youth can sometimes nearly lead to tragedy.*

That first time, I came to the Valley alone and at midnight on the cusp of winter. I was bent on storming El Cap, a teenage conquistador in search of pagan wilderness. My climbing career had begun scarcely three months earlier on a chalk-marked boulder. But my partner Joseph had declared the summit entirely attainable. I trusted in our combined imagination.

In my pack lay four coiled ropes, and in the belly of their coils, forty pounds of begged and borrowed pitons and steel carabiners. Because I didn't know Yosemite and because it was Christmas break and because I was stuffed with myth, a pair of red plastic snowshoes projected above the pack like a weak punchline. There were, of course, no deep drifts to breach. There was no snow to battle, no march to make, not so much as a hike to reach my destination. My thousand-mile hitchhike from Boulder ended quietly at a Conoco gas station, the Camp 4 landmark.

Mine was that classic entrance—in pitch darkness, no moon, no stars even—and I woke cold in the dirt to the Dream. Every pilgrim remembers it forever, that first view, that chalice of granite and sunbeams. I lay there, paralyzed by so much beauty rearing so high above the giant trees.

It seemed in all that light there could be no room for darkness.

Of course I *would* think that. I was eighteen. Tall, with a sunbleached mane long enough to pass for a Berkeley radical but with eyes much too innocent, I believed in everything back then. Spirits in trees. Infinite rock. Byron's edict—to light the sky like a comet, then take the snuff in full glory. I'd already vowed an early death. Joseph had made a similar vow. Before 30 there would be nothing left of us but echoes and Taoist poetry. Need I say we were students? Yes, college sophomores. We shared authors—Mishima, Kazantzakis, Borges—like hits of opium pitch.

Joseph arrived two days later with his girlfriend. He was a year older than I, she a year younger. He was nearly my height, with longer hair than mine, hand scars, and an infectious pantheism. He had once freed a panther from the San Diego zoo, he said. I didn't believe him. His girlfriend did.

The three of us set camp at the base of El Cap, beneath an overhang under what later became the Pacific Ocean Wall. It was cold and we kept a fire going. Our woodsmoke married the clouds at our feet. Until mid-morning when the sun melted off the mist, we got to pretend Eden belonged to us. There was nothing to see but the ocean of fog and the islands of stone. We were happily marooned.

This was in the days before boom boxes and satellite dishes, before there were lightbulbs in the john at Camp 4 (when it was still called Camp 4), before those daylong revolutions of tour buses with amplified voices. There wasn't even transistor radio reception, so the Valley was absolutely silent. You could think.

I made notes for a novel. Joseph prowled the forest. His girlfriend sang and painted watercolors. Every now and then sheets of ice loosened on the summit rim and the scree beds got strafed with glass rainbows.

On the fourth day we picked a line to the left of the Nose. The crack in the white granite leaked into heaven. Joseph said it was the Salathé and that ours would be the first winter ascent.

On the fifth morning, Joseph took me to one side. "I don't think we should climb today."

This was the third morning he had said it. I had to reply. Today I did. "We have to start the route," I said. "We should have begun days ago. We could be partway there already."

Joseph was adamant. He had the authority of more climbing and of "knowing" Yosemite. "This weather," he said offhandedly. And besides, he added, his girlfriend needed more time with him. Surely I understood.

I didn't argue, but my hunger was stark. As a sop, Joseph spent an hour showing me how to self-belay on aid. He wished me well. I started off alone.

It was a wild and woolly day. Bits of ice bombed down from the sunlit girdle on top. A gust of freezing updraft blew my own piss into my face. I placed the cold pins with aching fingers and drove them like tenpenny nails. The ropes whipped around at my feet with the agitation of foxes. I fell in love.

At the end of many, many hours, it seemed like I had risen halfway to the summit. The trees had grown small between my feet. The sun had wheeled around the Nose and turned the blue rock gold. In fact, I'd ascended barely 300 feet.

Near dusk I clipped my heavy racks of gear to the high point and rapped down. A hot meal was waiting for me in camp. Joseph's girlfriend fussed over my barked knuckles. She was awestruck by my soloing. So was I, now that my feet were back on the ground.

To my relief, Joseph now declared himself inspired. He confessed that the cold rock and hot love had started to erode his will to climb. But watching me battle partway to the sun, he had regained his desire. Furthermore, he'd talked to two other climbers hiking up to the base that day, and they had all agreed the route could conceivably go in winter.

It was settled then. First thing in the morning, Joseph and I would jug the fixed lines and leave the earth. His girlfriend would wait for us in Camp 4. We figured it would take five or six days to finish. We were so excited that Christmas Eve passed completely forgotten.

Long before dawn we were up and bustling. We wanted to get a jump on the short day and hit the ledges before nightfall. *Verglas* coated the scree—it was very cold. By headlamp, we packed our haulbag, cinching it tight. Joseph's girlfriend stayed in her sleeping bag as he kissed her goodbye. We headed off to our destiny, lugging the fat haulbag between us.

We promptly got lost in that inky darkness, or so it seemed. Slipping and struggling up the scree slope, we swept our lights back and forth across the rock for our green fixed rope. Finally, breathless, we accepted that we'd bypassed the rope.

We turned around and started down, hunting among the multitude of crack systems for our genesis. We descended all the way to the Nose without finding the rope. This was not auspicious. Mystified, we parked the haulbag and slowly clambered upslope again.

Not until dawn were we able to pinpoint the Salathé crack with complete certainty. All we could figure was that the wind had somehow deposited our rope out of sight. We backed into the trees, scanning the wall for it. There was the anchor . . . but no rope.

Our first reaction was relief and, for my own part, embarrassment. One of us could have been climbing on that rope when the anchor knots unraveled. I muttered something about getting my knots right the next time around.

But even as I made my apologies, Joseph erupted with curses. For above the anchor, our next rope was gone, and above that the final rope, too. At the highest point, where I had left all the hardware and our fourth rope, the rock was bare.

The truth crashed upon us like a wild animal.

We had been ripped off.

We stood there flatfooted. Our wings had been torn away, our odyssey squashed. There would be no ascent. There would be no trying. There would be no failing. This was something different than failure. I had journeyed to the furthest edge of my imagination, and someone had trespassed to shit on the Dream. For the first time in my life, I felt dirty.

Not one single piton belonged to me, not one inch of rope. It was bad enough that I would have to get a job in order to replace what we'd borrowed. Worse by far, Joseph identified our thieves as the two climbers he'd spoken with yesterday.

It was clear. In nearly a week, they were the only other climbers we'd encountered in the entire abandoned park, and obviously only climbers could have accomplished the theft. They alone knew of our plans. They had waited for me to rappel down for the night and then cleaned us out.

As we rampaged through the woods following bootprints in the frost, we found two carabiners wrapped with our red duct tape. At the roadside, the footprints vanished. "I never knew their names," Joseph said, "but I will never forget their faces."

Innocence is not something lost or given away. It is something taken. And you don't understand what it is until it's gone. What hurt the most was that the myth had betrayed me. The fraternity of ascent was nothing, after all.

Fifteen years passed. I traveled. I climbed. I loved. I lost. I loved again. Business brought me to California, and I pressed a rendezvous with Joseph, who was now in the film industry. Despite our separate vows to exit before the age of 30, we were both still alive and kicking.

We met at the base of El Cap for a day climb. Inevitably, for old time's sake, we selected the Salathé crack. While we shook out the rope, we paddled through a few memories of the camp beneath Pacific Ocean.

Nearly summer, the valley was jammed, the sun hot. To my eye, Joseph had changed very little, while I had changed entirely. The feeling was highlighted by this vast tombstone to my innocence.

Over the years, I had learned that life is filled with insult and outrage and deception. I had lost climbing partners to big mountains on the far side of the world. I knew women who had been raped and beaten, who, with their children, had been abandoned. On a lesser scale, I had spent months in Kathmandu jails for a fellow climber's smuggling and duplicity. Thieves—climbers—had twice more stolen gear from me in Yosemite. As a writer, I had seen my stories and articles plagiarized. Men in cars had spit on me and a bottle had been heaved at my bicycle spokes.

But even without those incidents—without the death and injustice and betrayals and ugliness—time itself had been revealed as a predator. My youth seemed to have been devoured. I had crow's feet now. My hair had thinned. My knees crackled and ached. And I no longer believed in the infinity of granite and sunbeams. There had been a time when my creativity knew no boundaries. Now it did.

It was a strangely bitter arithmetic that I applied to my past, strange because it didn't feel entirely deserved, stranger still because the afternoon was so glorious. I had never tallied up the injuries and trespasses like this, and wasn't pleased to be doing so now. Certainly I had many more triumphs and joys than losses. But Joseph's company at this place unlocked a door. It brought the ghosts swarming.

As it turned out, Joseph was thinking similar thoughts. For a few minutes we made dark jokes about the rip-off fifteen years ago, and what it had meant to us. Joseph said things had gotten so bad in the Valley that climbers had taken to rigging booby traps to protect their property. We agreed that no fate was too bad for a thief.

And then something occurred that was more extraordinary than the wildest fiction. It is my proof that this story is true.

We had just uncoiled the rope and were tightening our shoes, when a man walked up and asked to join us. His partner had bailed on him and he craved a few final pitches before driving home to the Bay Area. Would we mind a third in our party? He said he was strong and quick and wouldn't hold us up.

He had an Oriental face that looked Hawaiian. His shoes were worn, his hands scabbed with jam abrasions. He had his own rope. I glanced over at Joseph, who was oddly speechless. I took the initiative and said, "Sure, join us."

Without a word, Joseph cast a crazy look at the stranger and started up the crack. He seemed to attack it with his hands and feet. When I reached him at the top of the first pitch, he spoke. "Did you see him?" he said.

I looked down and watched the Hawaiian start off the ground. Just as the man had claimed, he was strong and quick.

"That's him," Joseph said.

"Him?"

"The thief. Our thief."

I exhaled my "No" as half a laugh.

"I'm serious." His frown pronounced it.

"Of course it's not him," I said.

The coincidence was too exquisite. Things like that didn't happen in real life. But Joseph insisted.

"There were two," Joseph said. "But I remember this one best because of his eyes. It's him all right."

"It can't be." I clung to my skepticism.

"I swear it," Joseph said. "I talked to him for an hour. He sat beside me right down there. It was cold in the shadows. We watched you solo. He was the one who asked all the questions."

"Impossible."

"I know," Joseph whispered. "But it's him."

"I'll ask him," I said.

Joseph shrugged. "He'll lie."

The more we talked—the closer the Hawaiian got—the more I accepted that destiny had just handed us our thief. The man's face was unique and the population of Polynesian climbers was small. He appeared to be our age.

We quit talking about the Hawaiian as he approached. He joined us at our stance, grinning despite our cold sobriety. I took the rack from Joseph and led on.

The climbing forced me to concentrate. It didn't allow any confusion. By the time I reached the next stance, I wasn't confused at all. I was angry. Joseph climbed quickly, as if the moves were more a distraction.

"What do we do now?" I said. He could see that I had converted. This was the bastard who had stolen from us. The same one who had broken my spell of innocence. And suddenly he was also the bastard who had raped and beaten my female friends, who had killed my partners, who had cost me again and again. Here was the vandal, the backroads terrorist, the lightning bolt, the violence waiting to happen. I had come to believe you never catch the bastard, but here he was.

"He fucked with us." Joseph's sunniness was in total eclipse. I had never seen him this way. Even his voice had changed.

"Now what?" I asked.

Joseph hesitated, then passed sentence. "I say he doesn't leave."

My silence forced him to clarify.

"We ask him to untie for a minute," Joseph said. "Then we kick him loose."

"Kill him?" I said. But my shock was less than authentic. "You can't kill him for stealing."

My caution only served to fan Joseph's fire. He was testing his own limits by testing mine, and seemed relieved by my answer. It allowed him to speak more graphically and violently, for now he knew his words were just words, not part of a murder plot. "We say he slipped," Joseph said. "His knot failed." He paused. "Fuck him."

Picturing the thief's face at the instant he would lose his balance and fall, I felt a radiant, barbaric pleasure. But then I pieced together the consequences. We would have to rappel into the blood and bones. For our single moment of revenge, we would have to live with a nightmare.

"No," I said. "No."

"What? Let him go?"

"Don't worry," I told Joseph. "I have a plan." By now the thief was at our feet, winded by marveling at the perfect day and the perfect crack. This granite was like a masterpiece, he remarked. We ignored him. I led up.

This was our final pitch. It was as high as I had climbed on Salathé on that cold blue day long ago. It was as high as this thief would ever go again.

Joseph swarmed up the crack. We stood side by side on the ledge and watched while the thief climbed toward our justice.

"Tell me," Joseph said.

"He stole our ascent," I said. "We steal his."

"Take his rope?" Joseph scoffed. "No way. I say we hang him by his neck."

Now it was my turn. In cold blood, I struck at Joseph's overheated fantasies. "We break his hands."

With a whisper, Joseph's voice lost its backbone. "His hands?"

"We take a rock. We smash his hands . . . by accident." I gave it the same pause Joseph had. "Fuck him."

Joseph turned pale beneath his sunburn. And suddenly, with all those years between, I understood why he had turned me loose on El Cap that winter morning. It had been to solo my energies away, to exhaust my ambitions. He had never intended to climb this wall with me. For the first time I realized that Joseph had never taken me seriously. In a sense my long-ago dream had been stolen even before our thief had manhandled it.

Now I declared my own revenge. It would not be Joseph's cartoon retribution. No more playacting. This would be real and bloody and filled with screaming and agony, and it would involve us—all of us—to our cores. I could hardly believe my own hard resolve. Yet, face to face with Joseph's charade of ascent fifteen years ago and with this thief's burglary of that, I could hardly deny myself the atonement.

Joseph was stunned.

"Is he the man or not?" I demanded, alive to Joseph's slightest dodge. I made this his responsibility.

Joseph took his time. His blue eyes lost their voltage. At last he nodded his head, slowly at first, then faster. "Yes. He's the man."

"That's that, then."

Grimly we cast around for a chunk of rock, heavy enough to drop on the thief's hands. I wanted to cripple the thief so that he could never climb or steal again. In Saudi Arabia criminals still get one hand lopped off in public squares. This was for real. It was for good.

But even as I rehearsed the deed in my mind, an unpleasant certainty hounded me. If Joseph could promote his fictions once, why not twice? No doubt he'd believed himself capable of a winter ascent, too. It was he who'd showed me that faith, like imagination, has limits.

"I'm going to ask him," I said. "I want to be sure."

Joseph straightened up. "It's him, I tell you."

"I'm going to ask him anyway."

"Don't. He'll just say it wasn't him. He'll blame his partner. He'll beg." Joseph was afraid. Was he afraid I would learn the truth? But that suggested he knew the truth. More likely, I decided, Joseph was afraid that if we confronted the thief, he would beg. We had gone beyond forgiveness, though, and Joseph knew it and was terrified.

It occurred to me that if I asked the thief, *then* dropped the rock, we lost our "accident." The mutilation would stand as what it was, punishment. So be it.

"I want him to know why it happened," I said. What I meant was that *I* wanted to know why it happened. Maybe I would even get an answer.

"Don't ask him," Joseph pleaded.

But I was going to.

We had selected a thirty-pound block with square edges for the job. Each of us hefted it. It was easy to see how it would go. The thief would place his hand on the ledge. One of us would distract him. The other would drop the rock flat on the thief's hand.

It all depended on timing. We had to trigger our assault the moment he gained the ledge. Too soon or too late and we would have lost our perfect revenge. If I wanted to have my answer, it had to come before the thief reached us.

As the Hawaiian worked his way up the final twenty feet, I talked to him. I complimented his footwork.

He modestly gave the credit to sticky rubber. I remarked on the piton scars—that fifteen years ago the scars had not been so deep. So ugly.

Joseph looked at me. The trap was set. We were caught now, all of us. There was no escape. The thief was just five feet away.

"That's what they say," the man replied. He was resting on his foot jams.

"It's true," I said.

"Oh, I believe you," he replied. "It's just I wouldn't know."

"No?"

"First time," he smiled up at us. Written on his face was a joy I suddenly recognized from long, long ago. "I'm a virgin. First trip to the Valley. I just started climbing four months ago."

"No." Joseph groaned it. For my part, a nausea took over, and I have had it ever since.

"Yeah, really." The man's pride was impeccable. It was as authentic as the stone we now laid back upon the ledge.

The stranger finished the crack. He got to his feet between us, exhilarated. "I've got to thank you. That was the best climb of my life."

Down at the base again, we coiled our ropes. The last thing the stranger did was shake our hands.

"You guys . . . " he paused. He became hesitant, a little bashful, even guilty. Here was his confession, I knew. We all have one.

"I'm new," he stammered. "But if you guys ever feel like climbing all the way . . . " We looked up at the golden summit.

Once upon a time I believed that in a chalice of light there could be no room for darkness. I was wrong. As the stranger disappeared into the forest waving his goodbye, as I glanced around for any last bits of gear, and the afternoon turned phantom, I saw the single thing that anchors my presence beneath the sun. Birds and angels may disown the darkness they cast far below.

But this one dark shadow is my own.

m0nkey on his rack

Nick Papa
July 1994

Dogs are frequently companions of climbers on the road—and they're often troublemakers. But even the most mischievious dog would be no match for Nick Papa's pet monkey, Fred. "For seven months, Fred and I climbed, traveled and lived together in my van," Papa writes. "It occurred to me Fred wasn't your average climbing parther, so I kept a log." The result is this bizarre and hilarious tale.

He boulders the intro moves and sets up below the roof. Conversation in the Gallery stops; belayers' heads turn. With that much muscle, abundant hair and no shoes, Fred is not your average climber.

True to form, Fred powers the move, slaps the lip and swings out one-handed. Matching casually on the quarter-inch edge, he brings a foot up. In fact, he brings both feet up, drops his hands and cops a rest. I stare transfixed, gaping enviously at my partner's ape index.

Feeding on the recognition, Fred reverses to a free hang, then busts an all-points-off lunge from a mantel to a stance on the same edge. Toes curled onto the edge, he thrutches wildly for a hold to keep him from peeling. There is none. Fred claws the face heroically, but the next incut is several inches up the overhanging scoop.

What Fred lacks in height, he makes up in strength. Dropping, he catches the lip and hangs from it again. It's not for encouragement that he looks to me now, but for a spot, the kind of spot only someone five times your height can give.

I step back.

Swinging desperately, he tosses his right toes into a pocket farther along the lip, then spans his arms, matching hand and foot in the pocket. Left leg and tail swinging over the void, he rocks up onto the pocket. Lurching off balance, he crosses his left hand from the edge to a sloper only a few inches above the pocket, then tumbles backward, dropping onto his feet on the slab below. Disgusted, he runs to the nearest person, jumps on him and bites.

Fred is an asshole.

■ ■ ■

For seven months last year, Fred and I climbed, traveled and lived together in my van. It occurred to me Fred wasn't your average climbing partner, so I kept a log to highlight both our cragging in the West and the depths of animal misbehavior.

February '91: Mount Nowhere

The hard white surface blurred under my skis, and from nowhere he came to block my path. I could never have guessed how much our chance meeting would change my life, nor did I have time to. In a sudden, violent over-correction, our collision was avoided and my knee shattered. Still in uniform, but at the end of my ski-instructing career, I cried the entire sled ride down.

May '92: Tempe, Arizona

Living in my van while visiting my friend James, I dislocated my knee again. A benchmark for change. It was while I was weak that Fred, really James' responsibility, was able to get close to me. Two weeks later we would leave to raft the Daly Canyon and head north for steep limestone. These were the moments of initial excitement. I did not yet know Fred well, and I could not have imagined what was to come.

June '92: American Fork, Utah

Fred is a two-and-one-half-year-old male Java macaque. He likes very much to be warm; in fact, he likes it hot. When the temperature drops, he will fight to get inside your shirt and liberate your body heat.

Hilary sees Fred shivering at the base of the Division Wall and offers to share bodily warmth, but Fred doesn't know how to share. He dives headfirst down her shirt and discovers a difference. Fred's very physical, like a bad first date. Squirming and flailing, Hilary screams for help. But I am belaying her husband John and can offer only advice. An already hilarious scene becomes even more so when the frenzied undulation attracts Burt the Wonder Dog, and John begins to laugh so hard he nearly greases off.

July '92: City of Rocks, Idaho

First of all, Fred eats anything and everything. When I say "eats anything," we're talking gum off the sidewalk, plastic, the rope, nails, rocks, dirt, bugs and chalk. He loves Endo so much that sometimes, instead of seconding routes behind me at the extension of his leash, he yards up it, sits on one of my gear loops and snacks from my chalk bag. Fred can store a significant amount of food in his lower cheeks, thereby securing your lunch much faster than if he had to swallow it. Which he eventually does.

The term "scared shitless" must have been coined when someone saw a monkey get gripped. Warming up on Tribal Boundaries, Fred makes the classic mistake—looking down. When fear strikes, Fred's intestines go into overdrive, and a hail of brown pellets ricochets down the slab onto an unsuspecting Jeremy, belaying us from below.

August '92: Rifle, Colorado

Fred stays clear of any ethical disputes by using his own strict code. He opposes hangdogging, often scoffing at me for blatant transgressions, for he, being an Old World monkey, cannot hang from his tail. Fred, who (as far as I can tell) never sweats, feels that chalk is only for eating and can often be seen at the base with a large chew in his cheek.

The last time I saw him use a rope was at Rifle. He was sketching up high on a new problem in the Firing Chamber when, pumped and desperate, he pitched off backward, spun in the air and caught himself on a convenient line, much to the dismay of the person leading above.

Fred prefers daring solo first ascents, and, not unlike boulderer John Sherman, abhors the use of cheater stones. You'll see Fred jumping again and again at tiny starting holds, each time falling to the deck undaunted. Fred may be the most powerful boulderer alive. Most of his sequences are so devious, thin and powerful that few of his problems will ever see seconds. To his credit, Fred is so casual on most desperates that he often stops to clean holds by licking the chalk off them.

September '92: Fremont Canyon, Wyoming

My friend Jeff has been traveling with us for a month now and has been very considerate of the limited space in my van. Fred does not share Jeff's kind disposition; his only concerns are for himself and his relationship with the "alpha monkey" (me).

So there's Jeff, eating his bagel, when suddenly Mr. Freddi grabs it out of his hands. Instinctively Jeff yells, "No Fred!" But Jeff is not the alpha monkey. In a transformation that rivals the Incredible Hulk's, Fred's forehead and ears triple in size, his jaw drops to expose gleaming white fangs, and his eyes fix like a gunsight. With an evil snarl, Fred becomes the hellion monkey and strikes the rest of our bagel stash with blinding speed. Dropping our last bagel and cream cheese into the dirt, Fred retires to the roof of the van with his spoils of dirt, cheese, pebbles and bread.

Jeff is steadily losing weight.

October '92: Smith Rock, Oregon

Sitting in the dirt at the base of Morning Glory Wall with Marcos Nuñez, Jason Karn and Freddi Monki, I suddenly realize I have been living in a dream. Perhaps the realization came when I flashed 11d or redpointed my first 12a. Maybe it was climbing with studs like Jim Thornburg, Martin Joisten, Rick Conover, Ed Barry and Steve Schneider that made me realize it. Or perhaps it is the combination of my aching tips, Fred's peaceful cooing from inside my sweatshirt and the beauty of the sunset over Asterisk Pass that makes me aware that my dreams have become reality. Whatever it is, it has taken six months on the road for it to hit home. For the first time in my life I have found peace, and it will give me a good perspective on the coming days.

One rest day at Smith Rock, I retire to my van for a snack. Sitting shotgun in Jason's bus, Marcos begins to laugh at something he's reading. Thinking he's talking to me, I open my window a little wider. Fred jumps out the van window, vaults through the bus window, grabs Marcos' hair and sticks his peter into Marcos' surprised mouth. I leap out of the van and the alarm begins to howl—earlier, I guess, Fred had snuck under the dash and flipped the switch that activates it. Fred jumps into the back of the VW, and I pull the sliding door off its track trying to open it. Marcos is cursing, my alarm is screaming, and I'm holding a Volkswagen bus door in both hands. I watch Fred run down the street.

Several minuets later I notice that Fred has returned to the van, eaten all the food I left out and left me dessert, a brown monkey turd, on the dash. The next day I take a 40-foot leader fall on Dreamin'.

November '92: Owens River Gorge, California

Having spent the first two weeks of the month resting, I was able to take a break from the daily rituals of van life and monkey maintenance, allowing me to look upon my partner's and my habits with fresh eyes. Each morning I wake before he does and push him out the window—if Fred wakes first, it makes both sides of the bed the wrong one to get up on. Next comes breakfast of either banana pancakes or oatmeal with raisins; Fred likes his heavy on the fruit, but too many raisins give Monki Bwei the pooey bum. "Monki Bwei" is his name when he is pleasant; "Fred" is what I yell in anger. The phrase "Bad Monkey!" holds so much terror for Fred that I need only say it once, and, raisins or not, his bowels let go.

Stretching follows breakfast, then the approach, some routes, lunch, a siesta, more routes, and we're back to the van for dinner. Fred spends his days pulling on boulders within the range of his leash. (I used to let him off the leash, but his raiding of people's packs, attacks on climbers and belayers, trundling, eating of guidebooks and my fear of liability put an end to his days of uninhibited curiosity.) Dinner of burritos or pasta is split almost evenly between us, as I approach my weight goal of 142 pounds, and he builds to 150 ounces. As soon as the sun drops behind the horizon, Fred begins a gentle purring, curled up against my digesting stomach. We share my sleeping bag and the farm dirt behind our ears.

December '92: Red Rocks, Nevada

In the same month that I redpoint a pair of 12bs and onsight as many 11ds, I sink to inexplicable depths. At first Fred and I are inseparable, but in less than two weeks I wish we never met.

Snow, which Fred has never seen, falls for a week straight, and rest days pass agonizingly, slowly chilling in my van. Fred's active urinary tract, combined with his hatred of cold, brings out a fury in me that even doing the routes I'm not doing couldn't have eased. More than a week passes without a climbable day, but I enjoy Christmas with my friend Jerry at the Holiday Wall. Climbing better, this day, than my early mentor, my world takes on a strange sense of circularity. Even Fred's pissing in my sleeping bag that morning might have been offset by my achievements in the afternoon. But I have grown to hate Fred, and the next day he makes me pay for it.

I drive to San Diego to visit my (then) girlfriend. With Fred under the seat, I'm not expecting a problem at the Yermo Agricultural Inspection Station. But fruit isn't all they're looking for, and, like the trouble he is, Fred rears his ugly head and catches the inspectors' attention. It's either kill Fred or drive back to Nevada. I don't like driving under the best of circumstances, and the thought of spending nine hours on the road only to end up where I started with nothing to show but a ticket makes me go ballistic. Fred's terrified face looks up at me from the passenger footwell. We split company on the morning of the 27th. Fred goes back to Tempe, from whence he came, to his old housemate James.

April '93: Tempe, Arizona

By late March it is too hot to climb in Hueco Tanks and, anxious to test my newly developed power, some friends and I drive to Datil, New Mexico, and the Enchanted Tower. The tower literally blossoms from the hillside, its summit three times larger than its base, and is covered with positive pockets. We are the only people there, I'm stronger than ever, and everyone is redpointing at their limits.

The night after I start work on my first 12d, it snows six wet inches. After barely getting my van out of its muddy lot, I drive down to James and Fred's place in Tempe.

Fred and I have been together for eight months and apart for three, but it's as if we never separated. On my last day in Arizona we go with James to Zoner and the Land of Nod in the Superstitions, where, while James and I are on Tone of the Bell, our friend Dean lets Fred off the leash. Not one to let opportunity pass, Fred hurries to my pack, opens the top, eats or destroys everything he finds inside, and then does the same with James'.

After rappeling down, we capture Fred and begin to lecture him. It's a crazy thing, the three of us screeching at one another, each in his own manner. I see that, not unlike humans, Fred must constantly push his limits, always testing to see how much he can get away with, how close to the edge he can stand, and how scared he can be and still survive. James, Fred and I are exactly alike. Each of us a climber, a primate and a freak—all of us brothers.

the zombie traverse

John Burbidge
March 1996

In this unusual story, a climber encounters the vagrants who live by a concrete wall, on which he wants to glue hand and foot holds for a bouldering traverse. Who owns such a wall and who determines what should happen on it? In many ways, the answers apply as much to our favorite crags of granite and sandstone as they do to forlorn concrete bridge abutments.

It was nothing more than an old cardboard refrigerator box flattened onto the dry riverbed rocks; still, the obvious indentation, human-shaped and pressed into the corrugation, tipped me off to the fact that some poor fellow was in bad need of a new Therm-a-Rest. The crumpled and muddy jeans balled up at one end of the box were evidence that the same guy could probably use a new pillow as well.

I stood there and marveled for a moment at the irony of how bad luck begets bad luck. Now that I was here, with my step ladder, PC-7 epoxy and putty knife, the very same guy who had been sleeping on the box was going to be in need of a new place to crash, because I was about to glue up a climbing wall in the middle of his bedroom.

It was the perfect bridge support for a glue-up; a 12-foot-high cement wall that stretched about 150 feet, located in the dry outer edge of the riverbed. The side that I wanted faced the water, hidden from the foot-and-bike path that followed the riverbank above. Because the wall was in the riverbed, I noted right off the bat that the possibility existed for the traverse to be almost completely submerged during high water; this seemed to add an appropriate natural dimension to the concrete personality of the structure. As I glanced over at the swiftly flowing water 25 yards away, I conjured up a disturbing image of myself being gushed off the wall in the middle of a fierce pump, too wasted to hang on as the water sucked at my legs, letting go . . . and heading for the Pacific.

Nice and discreet, I thought, looking around. No charges of vandalism for turning dead space into my own little training ground. Hundreds of smooth and juggy river rocks within an arm's reach, and killer late-afternoon sun to boot. I'd searched the entire town for such a spot. I knew I had arrived.

This dude's outta here, I thought, as I tossed the makeshift bedroll into the brush.

This western-Montana city was well known for its considerable transient population, especially in the summer, when numerous hobo heads could often be seen peeking over the boxcar tops as trains from Spokane and Seattle rumbled through town. I had heard of several popular hobo camps that existed beyond the outskirts of the city; this guy would just have to relocate. That's what being a transient was all about, anyway. They came and went and came and went, ragged faces turning into recognizable regulars around town. See them here and see them there, 20 times in one day and then not again for a month, until one day, if you happened to think about it, you maybe realized that a particular one hadn't been around for a while.

Outta here. Who knows to where. I shook my head and coughed out a short laugh. To think climbers called themselves bums. Climbers didn't even know the meaning of the word, next to these guys. These guys didn't pack around a couple of grand worth of gear; they were the real hardcores. They lived the life.

And this guy, I thought, tossing my stuff down right where the cardboard bed had been, this guy was going to have to live it somewhere else.

■ ■ ■

Two days and $30 worth of glue and duct tape later, things were beginning to take shape. I'd cleaned up about 20 pounds of broken glass and destroyed a couple of old firepits at the base of the wall, and, after collecting a big bucketful of likely looking river rocks and washing them in the warm, green water, I had begun to put up holds.

Huge, half-naked women, cartoon I-Dream-of-Jeannie types, and an oddly happy-looking giant skull smoking a stogie already adorned large areas of the wall, giving testament to the past efforts of some talented and twisted graffiti artists. For me, it was easy: Each eye got a handhold, each nipple and bellybutton a foothold. The skull accepted several amusing placements, including one on the only tooth in its smiling mouth; on any words of wisdom, I dotted the i's and put in periods and other stuff. Basically, the wall built itself. Like all truly classic routes, the natural line already existed; all it needed was a human instrument willing to put in the time and effort necessary so others could experience it. And while I was standing there on the ladder, immensely enjoying the fact that I was that human instrument, a voice spoke from behind me.

"You the artiste?"

Startled, I spun around and did kind of a falling jump off the ladder. As I picked myself up, I saw that a guy in a faded-green army coat was smiling at me, obviously a bum but, I decided, amiable enough.

"Is this your art?" he asked again. "Hey," he said, holding up his hands and showing me his palms, "don't worry. I won't mess with it." Slowly, he walked up to the wall and began fingering the rocks. "I won't mess with your art, or anybody else's." He looked at the ground as he spoke. "But then, you don't mess with mine, either."

I flipped off my Walkman and tried to wipe the sticky PC-7 from my fingers as I inspected this dude. He had bold, orange-ish hair and a freckled orange complexion, and slung over one shoulder he had an old, blue Boy Scout backpack. What he obviously didn't have was a clue. Apparently, he thought I was merely adding my two cents' worth of artistic expression to this slab of public cement.

"Because I got an important message to tell people," he was saying. "And I got these posters I'm gonna put up on the other side of the wall."

"Really," I said.

"Yeah," he said, looking at me expectantly. "And I don't want anybody messing with my stuff. And I won't mess with your stuff. Because this message I've got is important."

"Really," I said again, after a few seconds had elapsed. I gave my eyes a slight roll. "So. What's the message?"

His eyes lit up. "Raaah."

"Raaah?"

"Yeah. Raaah's the leader, man." He looked at me closely. "You know about the zombies, man?"

"The zombies?" I had to admit, he'd caught me off guard with that one. "Well, no," I said, "not really. What are they?"

The zombies, I thought. Jesus Christ.

The bum's orange eyes glazed over, and his face tightened into flaming, fused lines, and he began to rock back and forth slowly as he launched into his spiel. "Oh yeah, the zombies, I'll tell you right now, man, over 60 percent of Montanans are zombies already. They work in the back rooms of the supermarkets and restaurants, and that's where they put the worm eggs into people's food."

"Worm eggs?" I was getting into it now.

"The worm eggs get into people's brains, man! George Bush is a zombie!" He gave me his super-expectant look again, as if I should be terrified that big, important, former President George Bush is a zombie. "And all those other guys in the government, they're all zombies; they're running the country! Man, people don't know what's going on. Montanans just don't realize what's happening, but they're taking over here too, man; they've already been everywhere else. That's why I've got to put these posters up. To tell people."

"Well, dude," I said, trying to recover from the shock of having suddenly learned of this nasty zombie-takeover plan, "I guess I can buy the part about George Bush, but I don't know about Raaah and the worm eggs." I was beginning to get the feeling that I wanted this guy off my back, so, turning away, I donned my headphones and cranked the volume on Soul Asylum's "Made to be Broken." This guy could undoubtedly go on all day about this zombie stuff, and I had better things to do.

He stood there, though, behind me, and between songs I could hear him prophesizing in an increasingly alarming tone of voice. Finally, I flipped off the headphones and looked at him, just as he was turning to leave.

" . . . don't care, you're probably a fucking zombie too," he finished. Then he was gone, headed around to the other side of the bridge support, no doubt, to get his important message glued up, so all remaining "normal" humans in Montana—or at least those who happened to frequent the riverbank path—could be enlightened as to what was really going on.

Later that evening, however, when I was getting ready to leave, I saw him on his side of the wall, not gluing posters, but sitting in a makeshift chair of tires and plywood, opening a piece of foil that looked to have some sort of food in it, dumpster scraps or something. He moved his hands slowly over the shiny, greasy-looking meal and spoke softly for several minutes, rocking as he had before. Unbeknownst to him, I stood witness to his private ritual, the exorcising of worm eggs.

Jesus Christ, I breathed to myself, as I pedaled for home. This guy is bent.

■ ■ ■

The full magnitude of the truth behind this thought did not clearly reveal itself until the next day, when, sliding down the trail into the riverbed, I was suddenly confronted with not just a few weird posters, but at least 150 Xeroxed pieces of legal-sized paper covered with teenie-tiny, yet somehow perfectly legible, handwriting—an insanely encyclopedic history of Raaah and the evil doings of his race of zombies. It was all in order, all full of crazy but neatly designed charts and graphs, all of which were cross-referenced in a manner that made absolutely no sense at all. I'm no psychologist, but I'd managed to plow my way through Psych 101 in college, and to me it all appeared to be the looney, stream-of-consciousness-type thinking of a genuine paranoid schizophrenic.

According to the posters, the zombies were everywhere; they had invaded our society from every angle. They were responsible for the mysterious deaths of people ranging from Abraham Lincoln to Marilyn Monroe to Liberace. Place a finger anywhere on one of the posters and you were likely to come up with something like: "If your minister does not wear a cross, get one who does! All these are called 'major called ones'—and they are frequently copied (see section 'zombie monkeys' and 'zombie gorillas')." It was a garbled mess. It was out of control.

The orange guy was nowhere to be seen. I grew bored with the monumental task of reading the wall and went around to my side. True to his word, the guy hadn't glued any posters where I was working.

■ ■ ■

A couple of hours and about 10 holds later, I heard a strange noise coming from the other side of the wall. Walking around to investigate, I saw that a huge, hairy, transient-looking type was methodically ripping down the posters and throwing them to the ground. He was muttering angrily to himself.

"Say, there," I said, stepping up to the guy's side. "Say, man. Um . . . the guy who put those up asked me not to mess with them. You know?"

I stood there for a moment but was not obliged with an answer or even a sideways look . . . just more muttering. Great, I thought. Another goofball.

"Look, man," I said, moving a little closer. "I kind of made a promise to the guy who put those up that I wouldn't mess with them. What do you have to rip them down for? What difference does it make to you?"

The hairy bum turned to face me. He looked at me like he wanted to squash my head right there on the rocks. He looked at me like he had every reason in the world to rip these posters down, and all the reasons he needed to hate me at the same time. Finally, he turned back to the wall. "We don't want this guy in this town," he grumbled.

I was somewhat taken aback. The dude picked up where he left off with his angry muttering as he ripped and tore, and he didn't seem to notice at all when I backed away and slipped around to my side of the support. Christ, I thought, these guys are occupants of a completely different realm. Better to just let them fight their own battles; no reason for me to get involved to any degree whatsoever.

Later, however, I couldn't help but feel a twinge of guilt when the huge hairy guy came around the corner, dug the old piece of cardboard and the crusty jeans out of the bushes, and dragged them off into the evening. The deliberate manner with which he ignored my presence gave me the uneasy feeling that, like it or not, I was somehow involved with these guys and their crazy world.

But I blew it off. "Relocation, pal," I said aloud to the bum's back as he walked off. "That's what it's all about." I shook my head, tried to laugh and stared for a moment at the rock I held in my hand. I had a big jug smeared with epoxy, ready to go, but instead of taping it to the wall, I chucked it into the bushes and began to gather up my stuff. Suddenly, I felt as though I'd done enough for one day.

■ ■ ■

"You tore down my posters."

He was pissed. I climbed down from the step ladder and set the glue and tape on the ground off to the side. I held onto the putty knife.

"No I didn't, dude, but I was here when it happened. Some other guy ripped them down a couple of days ago. I told him not to."

The orange guy didn't believe me. "What'd he look like?" he snapped.

"I don't know," I snapped back. "He had a beard." This was getting out of hand. I was starting to get a little pissed myself, having to deal with this bullshit, and I decided it was time to tell this guy to get lost. I was just about to really lay into him when something about the way he was staring at my holds made me pull up short.

"Hey, look, buddy," I said, trying to be calm but forceful. Definitely forceful. "I didn't tear down your posters. I don't give a shit how many posters you put up. We have an agreement, remember? You don't mess with my art, and I won't mess with yours."

But he was gone. And a few days later, so were the majority of my holds.

■ ■ ■

The orange guy had become a presence around town. One of those bums that people chuckle at and maybe point out to visiting friends, remarking how transients lend color to this small Rocky Mountain city, and masking, perhaps, a somewhat embarrassing uneasiness that most people feel in the presence of bums and the uncertain circumstances that surround them. The transients often walked around the streets with a confrontational air, and they always seemed to be saying something, just jabbering, trying to push their loneliness onto you, because they had nobody else to talk to and nothing else to do. Most people, it seemed, carried around the feeling that a bum might, at any time, just decide to do something that was totally freaked out, something completely out of line, right in public, right in front of all the "normal" people and their kids, something so shocking that it might forever scar a "normal" mind. Why not? If you're a bum, you've got nothing to lose, so you might as well freak out, and, if possible, you might as well take somebody else, some "normal" person, down with you, just because you're a bum, and normal people are so easy to scare—they're so soft.

Over the past several weeks, the orange dude's prophesies and paranoias had sprung up on hundreds of telephone and power poles around town, and it began to seem like almost every time I drove somewhere I'd see him walking around, briskly, with his unique sense of purpose. The day after he discovered that his posters were ripped down, I saw him in the library, sitting at a four-person table with papers spread everywhere, busily scribbling away in his tiny handwriting, carefully piecing it all back together, diligently setting the record straight for an ungrateful and doomed world's benefit.

On that day, in the library, I'd felt a sort of grudging admiration for the guy's dedication to his task; two days later, however, surveying the smashed remains of a week's worth of my own hard work, I seethed inside and could almost taste the bum's orange blood. It looked as though he'd beaten on my holds with a pipe or a large rock or something, until either the holds had shattered or the bridge's concrete had simply pulled off.

As I stood there fuming, an orange head poked its way around the corner of the support and then quickly pulled back. Oh yeah, I thought, he knows what's up, all right. Furious, I ran around to the other side, his side, and grabbed him roughly by the back of his green jacket as he tried to scramble up the trail.

"What the fuck, you fucking ASSHOLE!" I screamed at him. "You knocked my fucking rocks off, didn't you? Didn't you?"

The guy cowered like a beaten dog. "I told you I didn't touch your goddamn posters, you crazy bastard! I told you I don't give a shit about your stupid posters!" I'd gotten caught up in my own tirade. With both arms, I pushed the guy backward and pinned his neck against the concrete wall.

"Listen man," I spat into his face, "if you touch any of my holds again, I'm gonna kill you. Got it?!" I threw him to the ground and had to stop myself from kicking him for added effect. He scuttled around in the dirt for a few seconds, his huge orange eyes twice as big as usual, then coughed and choked as he tried to get up. He looked at me sideways with fear and hatred. Boy, I thought, if this guy ever had any doubt as to whether or not I was a zombie, this just settled it.

"Well, screw him," I said aloud, back on my side of the wall, livid with rage at the prospect of spending another $20 on glue and tape and several more afternoons putting up rocks. "Fucking orange asshole got off easy."

■ ■ ■

I smelled smoke.

Stopping in mid-slide as I negotiated the loose dirt trail down into the riverbed, I held my nose up to the wind and sniffed like a dog, and at the same time, cocked an ear toward the wall. For a moment, I just stood there, still. . . .

It had been nearly three weeks since the completion of my glue-up, and, in that time, it had already seen a considerable amount of use. Various local climbers had learned of it through the grapevine, ever since the day I'd taken my roommate Mark and a couple of our buddies around to the wall and said, casually, "Hey, check this out." They'd had no idea; that was the way I'd wanted it.

So we'd started hanging out down at the wall after work or summer school, sitting by the river in the late evening sun, smoking a joint and maybe putting back a Foster's oil can or two, and, after a few days, people somehow got wind of things and began showing up on a regular basis. The colorful graffiti gave everybody a kick, as did the orange guy's zombie posters, and the reviews of my work came in mostly on the positive side, although Mark took it upon himself to lead the gang of critics who complained loudly about the "slippery, micro-turd worm moves" I'd slapped on the far end of the traverse. I told him I'd designed it with him in mind. Of course, when he finally figured out the moves his thinking began to come around, and, instead of bitching, he would spew massive beta at people as they pinched and paddled their way across what really was a sorry bunch of holds. Then, after they fell off, Mark would sigh and put on his slippers, and get up there and show everybody how to do it. The big bridge-support stud.

The bums had been making only rare appearances; usually they'd show up in the distance, back in the trees or behind the bushes on the sides of the riverbed, looking at us strangely, this new crowd of normal people who had suddenly decided they were going to hang out down here in a place that had never had normal people in it. They kept their distance; we kept ours; we ignored them, and they slept and lived somewhere else.

So it had been something of a surprise when Mark walked into our kitchen one day with his shoes and chalkbag, shirt off, sweat dripping, and proclaimed with finality, "Man. Those guys are tweaked!" Briefly, he related to me what had gone on that afternoon. Apparently, there'd been some kind of a confrontation between the orange guy and the hairy guy, Mark said, because when he'd arrived to do a workout, he'd found the orange guy lying next to the wall, "bleeding, and looking quite beat up."

"I asked him what was wrong," Mark had said, "but he just looked at me and said something like, he was being 'forced out' by 'them.' He said the zombies knew where he was, and they knew he was on to them, and they were going to kill him. I didn't see any zombies, but I did catch a glimpse of that big hairy guy ducking off into some bushes. I think he'd been watching me, so I just got the hell out of there."

At the time, I'd more or less blown it off. Who the hell cared what kind of problems these guys were having between each other? If the orange guy had gotten kicked around a bit, it was probably his own fault. After all, the hairy guy wasn't the only one who'd seen fit to knock some sense into the paranoid freaker. I'd had to do it once myself. . . .

Tonight, I was alone. It was almost twilight on a mountain-valley summer evening so beautiful that a bike ride and a quick pump seemed the perfect warmup to a night of quaffing beers. I'd thought I might run into another late-evening climber, or perhaps one of the fishermen who occasionally made their way along the river beneath the bridge, and now, sure enough, as I stood on the trail that led to the riverbed, I heard voices drifting up from the other side of the wall. But they were strange, guttural voices.

And I smelled smoke.

I crept slowly down the path and listened intently to the activity on the other side—my side—of the wall, and I heard the universally ominous sound of glass breaking. Quietly, I walked to the corner of the bridge support, and, after a moment's hesitation, I took a breath and peeked around the corner. In the approaching darkness of that summer evening, I was witness to what had most assuredly been going on underneath this bridge for years. It was a party. A party of bums. A party of bums, by bums, for bums. It was evil and out-of-control and spurred on by hunger. There was no one to hear them, no one to stop them. No one to tell them that they couldn't hold a rabid celebration of their hopeless lives, of the freedom and drifting anonymity and supreme next-to-nothingness they lived with every day.

A fire was burning against the wall, and several bulky silhouettes moved in and out of the light with quick, frantic motions. The hairy bum, the guy who'd ripped down the posters, was holding a huge boulder high above his head. He was poised to smash it down . . . right onto the foothold that marked the skull's front tooth. At the same moment the boulder descended upon the glued river rock and demolished it with indignant impunity, an empty wine bottle, tossed by somebody else, exploded high on the wall and showered the hairy guy and the surrounding rocks with shards of glass and splatters of wine. This screeching, simultaneous combustion of rock and glass sent the entire gang headlong into a frenzy of violent drunken howling and pushing and grabbing and wrestling. They played the way they lived—rough, expecting no sympathy and doling out none.

Thoroughly creeped out, I jerked my head back around the corner, lest I somehow become the unlucky object of the pack's destructive attention. My throat tightened, and one of my legs started shaking, as if perched on a dime edge, at the thought of being descended upon by them. Breathing hard, I grabbed my mountain bike and beat it the hell out of there. Behind me, I could hear it happening—the Zombie Traverse was being destroyed.

■ ■ ■

After several months, as fall and winter blew into Montana and my attention turned to skiing and road trips south, I came to realize that the paranoid fellow, the orange bum, wasn't around anymore. It seemed he'd left town, chased out, perhaps, by the other bums, by the climbers, by the zombies, or by whatever it was that was going on in his head. The posters he'd glued to the phone and power poles eventually got covered with other announcements, and, with no one to keep up the maintenance on the bridge wall, the great manifesto became tattered and yellowed beyond legibility. I never touched a single poster, this mostly due, I suppose, to an obligatory respect borne out of guilt for the orange guy, whom I'd mistakenly accused of messing with my art.

I never got the chance to apologize.

It was months before I went back. I stopped by on a winter's day as gray and thick as the concrete itself, a cold day, a day on which there could be no doubt that all transients had long since headed south.

The first thing I noticed when I walked around the corner was that some slob with a dripping roller full of red paint had obliterated the Jeannies and the skull; they were lost forever behind a millimeter of toxic spread. I walked around a little and felt out the few remaining holds, but I didn't even get off the ground. The loss of the graffiti conspired with the awful day to weigh me down. On such a day, the meaning of climbing, and all the reasons we have for leaving behind the flat old earth, was reduced to nothing. On such a day, the ground was good enough. For some, it was all they had.

I wondered all winter if I would muster up the motivation necessary to give the traverse another shot. Christ, the cost of glue and tape alone was enough to break a fellow who got by on a fairly modest budget. It was true that the transients were an inescapable and presumably valid part of this community; but shit, I thought, so am I. I was willing to exist in harmony with them, but the angry and surprised looks they shot us when we were hanging out at the wall were unmistakable evidence that they tended to view the whole thing as a battle for turf. I knew the bums didn't have a whole lot of discreet places left in a fast-growing town that was threatening to run them right over. I knew there were very few places within the city limits where they could still hang out, have a fire, drink some wine, and maybe roll out a cardboard box to sleep upon. Places where they are insiders—barely but crucially attached to the community of so-called normal humans. The bums need that attachment like we all need it.

Me? I had a nice home with carpeting and a shower, and I had a whole city I could roam around in and, just as important, blend into. I knew I could give the hairy guy and his buddies their one place back, their riverbed, this one single spot. I knew that the unselfish thing to do would be to disappear, like the orange guy, the one person who'd wanted everybody to live in peace and leave one another's art alone. I could disappear from their world, leave it behind, and go pull plastic at the university every night with all the gumbies. But I didn't fit in with that crowd and I didn't want to. I had that much in common with the bums—and I was willing to fight for the broken glass and graffiti world underneath the bridge. Like the bums, I felt comfortable there.

I picked up my checkbook one warm spring day and headed for the hardware store to stock up on PC-7, putty knives, and duct tape. As I pedaled, it occurred to me that I could use a few allies this time—another battle was about to begin. I decided to pick up a few friends along the way and grab some paint at the store. Lots of colors, we'd have a party of our own down there as we redecorated the wall with pictures of huge Jeannie ladies and cigar-smoking skulls. And holds, lots and lots of smooth river-rock holds. The Zombie Traverse was coming back to life.

a day alone

Barry Blanchard
November 1997

What goes through the mind of a climber who free-solos a towering pillar of ice? Most of us will never know, but this piece offers a glimpse of one mind at work. Barry Blanchard has been among Canada's foremost alpinists for more than a decade. Is this fiction or autobiography? He doesn't say.

He slept in the back of his pickup, with the windows cracked one quarter of an inch and two candles burning on the small shelf of the camper shell. The candles warmed the camper, but they were too small to burn all night. The flames shrank until they were small blue spheres that sat lightly on the blackened wicks, then they sank into the pools of wax like fragile flowers sinking into water, small lights conceding to the darkness. The candles died within minutes of each other.

Outside, the ice on the reservoir thickened and groaned. Like broken slate, plates of ice uplifted along the shoreline, and every hour the ice cracked in the arctic air with the clarity of a rifle shot.

He woke to darkness and cold. The tip of his nose and his cheeks were numb, as if they'd touched algid metal for the hours he had slept. He pulled his face back into the sleeping bag, slid his gloved hands up to his face, and flooded the vault of his hands with warm breath until he felt the blood and life return to his face. Then he pushed his head outside the bag. The cold slapped him, and he felt a heavy, lethargic urge to retreat into the dark heat of the bag.

A beat passed. He wrestled a determined hand out of the bag and twisted on his headlamp where it lay snug over his toque. The light reflected silver from the frozen breath that glazed the portal of his bag. His frozen breath. He saw the death pools of the candles puddled on the sill of the camper.

An aloneness rose in his throat and with it an instant of sadness that compressed lines into his brow. He allowed the sadness to sweep him beyond loneliness to a place where he saw he had left the planet and was slowly rotating in the black void of space, arms and legs outstretched in a star-shape, his image shrinking. The sadness scared him then, and he clenched his eyes and hauled himself back to his resolve to climb.

He found the bag's zipper and opened it in jerks. Raw cold bit through his liner gloves, and he cupped his hands to his mouth and blew warm breath into them, then scraped around the open bag to find his outer gloves. He had slept in his underwear and pile.

He pivoted off the tailgate. The hair inside his nostrils bristled with frost; the skin on his face, thighs and groin contracted in horripilation. He spun, heaved up the tailgate, and eased the canopy window down. He always expected it to shatter in this kind of cold.

Brittle snow crunched under his boot soles like deep gravel. He paced outside as the truck idled to warmth, throwing his arms through full circles to force blood down to his hands. Shadows stretched from his feet and lost themselves in the reach of the headlights. Surface hoar sparkled along the light's edge like a border of jewels, or, he thought, like the last stars that stand guard between living space and the void.

Slowly, wet stains spread up the frozen windshield. When the stains had grown to eye level, he climbed into the pickup and coaxed it into gear and pulled onto the road.

■ ■ ■

He crossed the Bow River at the Stoney Reserve. The dark forms of immobile vehicles and solemn, unlit Indian shacks passed by his window like slowly grazing bison, and he thought how his people—the *métis*—had once been buffalo hunters. He imagined his grandfather hunting in winter in the Qu'Appelle Valley, 70 years after the buffalo had gone: galoshed, shin-high moccasins punching holes into the brittle snowpack or trenching furrows through the waves of drift that clung to the lees of the coulees; knot-heads of ice clinging to the rough, home-tanned leather and growing to the size of marbles through the day. He recalled the photograph of his grandfather going to war in the uniform of the South Saskatchewan Regiment, and he considered how men dressed in cotton and wool and leather had fought in wars and worked in the bush and even climbed mountains. Pressing his hand over his thigh, he felt the slickness of his pile salopettes and how they slipped over the silken underwear—not silk, but synthetic material made to be silken. Modern clothing gave him an advantage when he was climbing, yet he preferred to dress at other times in cotton and wool and leather boots or moccasins. He felt closer to his grandfather that way.

He drove in silence. The smooth and efficient hiss of the truck was muted by his imaginings.

Breaking from the forest at the top of the big hill, he saw the full frontal escarpment of the Rockies, and the whole of it—the forests footing it, its borders of snow and the sky above—was immersed in the shallow, blue light of brumal dawn. He pressed the pickup into four-wheel drive and braced his hands against the steering wheel as the truck walked unevenly down the steep, rocky grade. The pickup always felt like a horse to him here.

He stepped out of the pickup at the national-park boundary, a saw-cut three meters wide up both flanks of the valley, as high as there were trees. Climbing boots and outerwear on, he started down an old road to the Ghost Lakes.

■ ■ ■

Dominant west winds, hauled earthward by gravity and squeezed between the shoulders of the mountains, had raked hard at the dehydrated and cracked bed of the second Ghost Lake. The wind pried loose grains of dirt and bounded them over the lake bed to deposit in the lee of the lake's eastern shoreline. This was the only place in Alberta he had seen dunes. Scythe-curves of dirty snow clung to the dunes and extended their taut lines—lines of wind visible on the land, as abrupt and clean as a saber cut, as beautiful as the sickle moon.

The melding of color from earth to snow was so perfect that he could not differentiate between the two, until he punched in knee-deep and saw the pure white basements of his footfalls. He strode down the dunes' western flanks with long, scree-running steps. He passed stunted and leaf-bare, trembling aspens, their trunks engulfed by the dunes and branches dusted dirty-gray by the wind. A branch clawed at his upper arm and snapped against his passing, the core alive and yellow but frozen brittle, the skin like brindled bone. Coyote tracks pressed into the dust of the lake bed and in the cement-colored snow that bordered it. The coyote's crossings seemed haywire, erratic, purposeless.

It hadn't blown since the invasion of the arctic air mass, maybe longer.

■ ■ ■

He first heard—then saw—the great wind. A distant rumble reverberated against his sternum; a cloud rode the surface of the lake. The cloud advanced, and the rumble exploded to a towering roar, and he saw trees upslope bend crazily and erupt in plumes of dust. One tree sparred with a loud crack, and its top was hauled away; the fractured trunk snapped back into the wind, its core white and jagged and swaying.

Then he was smacked full-front by a screaming wall that bowled him off his feet and thumped him down hard into the dirty snow. The abruptness of the blow defeated his reflexes, and he fell back-first. The breath burst out of his lungs, and dirt and ice shards hacked at his face and stuffed into his throat. Panicking for air, he rolled over and hauled his arms to his head. He coughed and swore and struggled to rise. The chinook jerked him like a puppet, and he went to his knees twice before he finally got to his feet, five meters away.

Immediately, he sensed the warmth, like a biblical flood of Mediterranean water. His face grew slimy with slush, and the breast-plate of frost he had grown since leaving the truck cracked and began to fall away. The hood he'd worn all morning suddenly felt claustrophobic. He ripped it off and faced the wind. Water teared across his face and pooled in his ears, and he pulled the toque from his head and laughed and leaned his shoulder into the slapping hand of the chinook.

■ ■ ■

Reaching the climb was an act of will. He continued against the chinook because he didn't know if he could. The wind pummeled him constantly; each footfall was a thrust for balance. Leaning heavily on his ski poles, arms fully extended like outriggers, he felt as if he were wading a thigh-deep river, then the wind would back off, and he would stagger and fall into the void where the wind had been. He wondered if, by continuing, he was expressing courage or arrogance or stupidity. Finally, he decided he continued because it was daytime and therefore his time to move. By the time he reached the ice, he had fought his way out of his pile jacket and vest. The wind pulled at them like a desperate thief as he stuffed them into his pack.

The first plant of his ice axe released a starburst of ice shards, and the wind swooped them into the vortex in front of his face. He feared he would take a piece in the eye and shouted to himself, "No! Not in the eyes, Pelletier. You can't take it in the eyes!"

He stepped down and unslotted the tool, pulled goggles from the top flap of the pack, and turned into the wind to put them on. Their rose tint gave the sun a beautiful halo and outlined the clouds with a prism of oily blues and reds. High above, he saw the tidal edge of the chinook, a three-layered standing wave of cloud that ran south to the horizon in defined arcs and troughs.

He climbed. The chinook harassed. It pushed him hard one way, then oscillated to slap at his opposite flank. There was an eerie instant of calm between the changes, as if the wind were deciding where to land the next blow. He told himself to stay wide and stable as he climbed, that the wind couldn't take him off that way. A lie, but one that gave him hope.

The waterfall wore a façade of icicles the size and shape of church candles. They hung like wind chimes between finned roofs of ice and the flat floors beneath them. The icicles gave no support. When he cleared them with the head of his axe, they tumbled toward his face or piled on the small ice floors like toppled columns. Some bounded off his second tool, whose shaft vibrated like a tuning fork. His forearms were tiring, and fear now heated his climbing and pushed on him to move.

Then the chinook backed off completely. He felt his wind-borne weight sag onto his locked left arm, and he knew he was about to be hit hard. He swung his right tool in desperation. The head of it shattered icicles and swung into the useless space behind. Then the booming wind hit him. It grabbed him from the inside and spun him through a half-rotation, so he faced out from the mountain with both feet and his right tool clawing uselessly at the air, like an insect turned over and struggling. "Fuck!" he screamed.

The wind let up, and, in blind animal fear, he kipped back to face the ice. Body fully extended, he hung from his left tool. Reflexively, he kicked and hacked for purchase. The right crampon stuck, the left clawed and skated blindly. He bludgeoned the right tool at the icicles, and, on the third flail, he felt it stick. But the ice cracked deeply, and the fracture radiated to his left tool.

Fear. Far beyond the fear of falling now—this was the man's hard-wired imperative to survive. The fear flooded his blood with fire; the command to flee shouted out at the cellular level. He shook in a vibrant and clear tremor.

With quivering thrusts from the heel of his hand, he coaxed the left tool out of the ice. His left foot swung behind him, and his crampon points tapped the ice in a desperate Morse code. He swung and swung the left tool. It fishtailed in his hand. He cried. He swung again. The pick sunk. He heaved his left foot back and kicked it into the ice. Tears ran in small streams from the corners of his eyes and down the lines of desperation and regret furrowed into his face.

Gently he lifted the right tool. It unhooked with no resistance, but the hammer caught on the icicles, and, when he pulled on it, the cracked slab separated with a sick clunk. The slab hit him in the chest and pushed a prayer from him—"Please, God, no!"—then grated past him and thumped a crater into the snow hundreds of feet below.

Catching the right tool in his hand, he swung, and perhaps providence guided the tool to a solid stick. Desperately he thrashed to find a sling and clip into it, then he fell onto the security of his harness. Breath raged from him. He shook violently. His tears flowed freely, and the chinook smeared them across his face.

It had been one minute and twelve seconds since the wind hit him.

Fits of shaking seized him with the regularity of a woman's labor. The intensity slowly waned until he was limp and spent, a man destroyed. He raised his head slowly, and began to go about the tasks of getting down. Easing onto rappel, he took a last look at the anchors. He laughed and shouted into the wind, "Oh yeah! I'll be backin' off!"

■ ■ ■

He cut out of the drainage and onto the exposed grass and low growth of the south-facing slopes. The ground gave softly beneath his boot edges. The chinook had scraped the main valley down to nude brown earth.

He found the ram's skull halfway down to the main valley. The ram had been old, his horns curling nearly to full circle, the husks grated from the weathered bone beneath and pressed into the ground. A vacant cornucopia. He strapped the ram's skull under the lid of his pack and continued down.

Two hours later, he walked into the meadow where his pickup was parked. Shallow pools of water had collected. The chinook, now blowing firmly but without violence, corrugated the water's surface.

He cooked on the tailgate in the lee of the pickup, collecting water from shallow pools with the lid to his pot. The ram's spirit watched through an empty eye socket from the shelf of the camper, beside two pools of candle wax. That night, the man would sleep again in the back of the pickup, and, perhaps, the next day he would climb.

night and day

Christian Beckwith
January 1998

The Soviet Union's rigid, hierarchical climbing system produced some of the finest mountaineers in history, yet the collapse of the Communist government prematurely ended many of their careers. In this story, American Alpine Journal editor Christian Beckwith unveils the old system and offers a moving and sympathetic look at the climbers that history passed by.

Bishkek. The ache of traveling knees, the smell of unwashed crotches, thin films of sweat in dried waves on our skin. An English show dubbed in Russian blares from the television. Fetid water gurgles in canals along the streets. Old women in dirty flowered dresses sell laundry soap and Fanta from rickety wooden tables, while Ladas, the vehicles of the proletariat, fart foul exhaust into thin blue air. White mountains float against the haze. We have come to climb them. What we do not expect is the struggle at their feet.

In Bishkek, the capital of Kyrgyzstan, as in all the former Soviet Union, communism is finished, ripped from the body politic by a violent and imprecise event called perestroika. Grandparents reel from it, unable to comprehend what it has done to their world. Their grandchildren will feel its effects for generations to come.

We stay with Vladimir Komissarov, the president of the Kyrgyz Mountaineering Federation, in a disheveled but clean bungalow that serves as the offices for his guiding company. We drink vodka. Everyone always drinks vodka.

"Before perestroika," Vladimir tells us, "there were thousands of climbers in Bishkek. Now there are very few. Before, all you had to do was show up at the alpine center for gear provided by the government and go off into the hills and climb. It was very easy. Now everyone worries about work; there are few people who can afford to climb anymore."

We climb for a week in the Ala Archa National Park, 4,000 meters above the city. We meet two Poles, three Ukrainians and a German. The last three days are spent in a climber's hut in solitude. It's late in the season, and the weather is changing, growing colder; a day after our first climb, it snows half a foot in twelve hours. Is this why there are no climbers here? Maybe. But with a city of more than 600,000 people six hours' hike away, we have our doubts.

■ ■ ■

Brady Van Matre and I skittle across the Ak-Sai glacier in the half-light of dawn. I start first, trailing a rope and moving quickly, then carefully, as the surface crust becomes broken by sporadic patches of ice. Hundreds of feet up, a tumble of large granite blocks pokes out of the couloir. I size up the remaining distance and concentrate on my placements.

Ten feet from the boulders, the snow goes to hell; my tools plunge into depth hoar before finding purchase. Suddenly my foot hits a rock the size of my chest, and it sags away from the slope.

"Brady! I'm holding a huge rock in with my foot! Traverse right! Get out of the way!"

"Hold it in!" he yells, his frightened face wide-eyed beyond my heel. I hold the rock in the snow with my frontpoints, then, when he is sheltered from sight, gingerly relax. The rock holds. I climb above it and angle toward the outcropping for an anchor. When Brady reaches the loose block, he taps it with his pick. It rolls slowly from its hold and rushes down the mountain.

With Brady in the lead, we move onto the first real ice of the climb. When the ropes go tight, we climb together, both ropes singing up straight as sunshine, our tools sinking snugly two inches deep.

I am out of breath by the time I reach him. To our left, the ice surges through a vertical bulge.

"How's the anchor?

"All right. Decent ice."

The ice is brittle here, steeper than anything I've ever been on in the mountains. I sink a screw 15 feet out. In some other part of my mind I hear Brady mumbling words of encouragement. At a block I put in another screw. That leaves two for the next 150 feet, plus the belay. The ice is thin and hollow, melted out by hidden rocks that flatten my picks. Confidence recedes. I am too shot to continue. I downclimb to give Brady a try.

He climbs to the right of my line and gets in one screw 20 feet above me and another 80 feet above that. Then he disappears from sight, and I feed the rope out without break for half an hour. When I start climbing, I find that his second screw was his last.

Aiiy! Calves burning, lungs cracking, I thunk-thunk my way up the couloir. The sun has burst over a corniced ridge; sweat percolates up through skin, soaking polypro and Gore-Tex. Brady sits comfortably on a golden granite bench. My sweat mingles with my hair and drips down behind my glasses.

"Nice lead," I mumble.

"What did you say the highest you'd been before was?"

"Thirteen seven-twenty. What are we at now?"

"Ten feet shy of that," he says, holding out his watch. The couloir rises along a rib of rock for another 200 feet, then flows around a gendarme in the middle. We have to climb left; a healthy serac caps the ridge crest to the right.

"All right. Give me a shot of water, and I'll get out of here."

The climb exits via a steep constriction that hugs the rock wall of the gendarme. A nut in a fissure, then away, up into the ice as it gets steeper and narrows, picks sinking perfectly, angle kicking back, rock wall looming behind my head. Up, up, each placement bringing me that much closer to the two-foot gap at the top, until finally I am swinging awkwardly into névé and dragging my gasping body into bright sunshine and flat snow. The wind howls and whistles; I stand blinking in the bright glare, awed at the panorama of mountains before me. I throw down my tools, then my gloves, sling a flake, and bring Brady home.

■ ■ ■

A few days later, we take dinner with Vladimir at a hole-in-the-wall restaurant 100 yards from his office on Panfilov Street. We drink chai, the dark brown tea leaves congregating in swirling clumps at the bottoms of our glasses, while he questions us about our impressions of the Ala Archa. We tell him about the couloir.

Vladimir, who was born in Bishkek and has climbed in the Ala Archa all his life, squints his blue eyes in concentration.

"That route is new route," he announces, sliding skewered pieces of mutton and raw onion into his mustached mouth.

We find out later it wasn't—it was climbed in 1984 by a Yugoslavian team. But there, in the restaurant, we whoop, high-five, spit food onto the dirt floor. Impossible! How could that be—such an obvious ice route, and in mountains 40 kilometers from the nation's capital? Something here is amiss, and we set to work trying to understand how such stunning mountains could be so free from climbers, how obvious lines could be thought unclimbed, and how the rest of the world has not heard of these wild knotted regions in the lost heart of Asia.

■ ■ ■

In 1985, when Mikhail Gorbachev took over the reins of the Soviet Union and began the ambitious and unprecedented campaigns of *glasnost* (openness) and *perestroika* (restructuring), he pitted his resources against the biggest bureaucratic monster in the world. For six years, he whittled away at the inherent flaws in its defenses, encouraging movement toward democracy, private enterprise and self-will. In 1991, the monster thundered to the ground. The Soviet Union was dead.

In the West, Gorbachev was a hero. At home, however, he had destroyed a safety net that provided security to his people. The basic services they had depended on—medicine, education, housing, even jobs—were gone. The money to fund the services had all been used up in the arms race with the West. What was left fell into corrupt corridors of power. People who had spent their whole lives paying into a communist system were told that the system was bankrupt, and that now they would have to fend for themselves.

Vladimir and I eat again at a restaurant a block from Victory Square. I pepper him with questions, trying to understand what has become of his country, its people and its climbing.

Within the framework of communism, he tells me, alpinism played out along a party line. Money was available to the "trade unions" of the USSR for various pursuits, such as mountaineering. Climbers had to join a union, which gave them vouchers to attend mountaineering camps and equipped them with all the necessary gear.

Every climber in the USSR had a Climber's Book, into which was entered a complete history of his or her climbing, validated by the appropriate officials. All climbs in the Soviet Union had a grade. If a climber wished to climb a route of a certain grade, he or she had to receive permission for that route from the appropriate alpine officials. In order to receive permission to climb, say, a 3b, the Climber's Book would have to be stamped with satisfactory performances on routes of 3a.

"But Vladimir," I ask, trying to understand, "what about new routes? How could you explore new areas and new mountains if everything had to be preapproved?"

"You had to make a proposal to the trade union," he says, "detailing where the route would be, how hard you expected it to be, and showing your qualifications to do such a route. Then, if you were given approval, you could do the route. Once you had permission, you climbed—because, if you didn't, for weather or any other reason, it went down in your book, you were demoted, and you had to start again at a lower level in order to regain your status."

I ask if there were any climbers who refused to take part in the system, who climbed independently, rebels in a lockstep world.

"Impossible," Vladimir says. "To do that you had to have your own equipment, which meant you had to have your own money—and nobody had that kind of money. To climb, you had to climb with the trade unions—and to climb with the trade unions, you had to climb by the rules."

I had read of climbing competitions within the former Soviet Union—speed ascents on the limestone walls of the Crimea and the 7,000-meter peaks of the Pamir and Tien Shan, and new-route competitions in the Caucasus and Pamir-Alai. Many Western climbers denigrated these competitions, and I, too, felt they ran counter to the spirit of climbing. Now, I began to see that competitions were an organic extension of a system where climb and climber—as with every other member of society—were graded according to strict rules.

But this was the world before perestroika, and, just as the complete order of Soviet life disintegrated in 1991, so, too, did the internal scaffolding of the alpine system collapse with Gorbachev's historic moves.

■ ■ ■

A few days after our return to Bishkek, Brady and I hook up with Matthias, a German we'd met in the Ala Archa, and the three of us journey to the Karavshin region of the Pamir-Alai. After a three-hour, surprisingly uneventful flight to Osh, the second-biggest city in the country, we hire a taxi to take us to the Tajik town of Vahrook. From there, we make the 50-kilometer approach into the Ak-Su Valley in two days.

A helicopter arrives two days after we do and disgorges a load of guided Brits and our food and gear. The next day, we climb a 1,000-foot pillar beneath the massive granite tower of Pik Slesova (a.k.a. Russian Tower). Brady calls down from the first belay that he has found bolts at the anchors—Simond, shiny and new. We use them all the way up and all the way down.

The British tents spiral in an irregular circle around the mess tent. There are 12 clients and three guides. Two Kyrgyz men, Yuri and Agnar, erect a comfortable hamlet that includes a shower, a privy (complete with an inexhaustible supply of toilet paper—the first we've seen in Kyrgyzstan), a cooking tent, a food tent, a deep fire pit and half the dead wood in the area to feed it. The group eats lavishly, three times a day. A woman named Ala works from five in the morning until ten or twelve at night to prepare their food. We quickly make her acquaintance.

She is petite, with fine red hair and slightly Kyrgyz eyes. I ask her the morning after our climb whether she is a climber.

"Oh, not now, not as much. One time, yes. But, since perestroika, I can only work."

Ala is a doctor by training and profession, but the pay of roughly $20 a month is not enough to make ends meet. When she found an opportunity to work as an interpreter for a British company, she quickly parted ways with medicine. She has a month off a year, and, for two years in a row, she has gone to the mountains with Pat Littlejohn, the leader of the

British party, who pays her the equivalent of a doctor's monthly salary every day. I talk to her about her climbing.

"I was part of a team in Bishkek," she says. "There were six of us, and we only climbed together. My job was to climb quickly, to never hold up the team. We climbed well—up to the 5b level. I jumared the pitches—I never led—but it was very fun. We were a good team. Every summer, we would climb two times together, for two weeks in May and two weeks in November, at the alpinist training camps. The trade union would give us permission to leave work for three weeks and pay 30 to 50 percent of the fees. Everyone went to the Ala Archa to climb. But after perestroika we stopped climbing—the trade unions were dissolved, so there was no equipment, and we all had to work too much to climb anymore."

I ask her how perestroika has affected climbing in Bishkek.

"Before perestroika there were many, many climbers—not less than a thousand. Not everyone who started in the camps continued climbing, of course—maybe 70 percent kept climbing, and of them 30 percent were serious climbers. But now there are maybe 20 or 30 climbers in Bishkek—no more."

We attempt two more peaks in the Ak-Su and fail on both. During the days, the thundering granite walls keep the sunlight from the valley for all but a few hours; at night, their silhouettes block out the stars. We kill a sheep in basecamp, Vladimir skewering the meat with slivers of wood, the blood running into our hands as we eat. We climb one last line, then drive a train of pack mules down the valley to Vahrook.

■ ■ ■

Back in Bishkek, I speak with Alexander Agaphonav, a master of sport, the highest level one could achieve in Soviet alpinism. He is 38, 5-foot-8, and compact, with legs that explode when he walks; he wears torn and patched denim shorts gone thin over countless summers and innumerable climbs. His blond hair is thin, and his wide Russian face is punctuated by the lumped bridge of a nose broken long ago. While a woman from a local trekking firm translates, he talks about climbing then and now.

"We had two kinds of climbers before perestroika: one professional, the other who climbed as a lifestyle. The professionals were part of 16 teams in the former Soviet Union. Each team had eight members. Of these, four were masters of sport, and four were apprentice masters. The masters were masters of six classes of climbing, for which there were competitions: rock class, for mountains not higher than 5,000 meters; technical class, which was climbed regardless of height of mountains; ice class; high technical, on mountains higher than 6,500 meters; traverse class; and winter class. Sometimes there were competitions between the two kinds of climbers in three different classifications: high peaks, middle peaks and low peaks. The winners became part of the professional teams.

"The professional climbers received a salary from the state—though it wasn't enough to live on, and the climbers had to find other ways to earn money. Meanwhile, you had to always train, always be in shape for the competitions, and you always had to take part in higher levels of competitions. It was a special kind of life, not a sport—you had to live to climb. It was a very hard system to take part in."

At his apartment, Alexander shows me his Climber's Book. In it are recorded and duly stamped all the climbs of his career, from his earliest ascents in the Ural Mountains, where he had lived before perestroika, to his climbs as a master. He pulls a gray notebook from a shelf and opens it. Pictures of a 3,000-foot wall are marked with lines and numbers—the record of a first ascent in Tajikistan. He explains that such detailed itineraries are required of all first ascents in the former Soviet Union. Three pages of text come after the two initial photos showing the peak, then more black-and-white photos of the climbers themselves. He points to a dot high above the belay, a climber nailing an incipient crack on a 20-foot roof in a blizzard.

"Me," he says.

Before perestroika, Alexander traveled every two weeks with his team around the mountains of the USSR. After perestroika, the professional teams collapsed. He moved to Bishkek to be close to the mountains, but he struggles to make ends meet. He is most qualified to teach climbing, but now, of course, "The trade unions are broken up, and climbers haven't the money to pay for trainers without them. Sometimes, I train young people for no money, but it is very hard. No one helps us to go climbing, and we don't have money to go on our own."

His words echo everywhere. There is no money. There is no work. There is no climbing. The reality of post-perestroika alpinism is the same as life everywhere in the former Soviet Union: a stripped-down version where children in their mid-20s support their parents on salaries that wouldn't afford a pair of jeans in the States, where the only common thing left from communism is poverty, and that asphyxiating blanket affects all, young and old, smart and simple, bus driver, student, factory worker, climber.

I remember what Vladimir told me one night over too many shôts of vodka, in the basecamp tent of the British, beneath the brilliant stars of Central Asia. He said that the former Soviet mountaineers, in their prime, were perhaps the best in the world. Once they had overcome the bureaucratic hurdles necessary to attempt a route, nothing could dissuade them—not winter, not weather, not lousy gear nor jumping nerves nor any of the excuses that crop up in the West.

Within the walls of communism, Soviet mountaineers ascended stunning routes on tremendous mountains in total seclusion. They were the best-trained mountaineers in the world, and, in 1982, when the first Soviet expedition visited the Himalaya, it successfully climbed a new route on the Southwest Face of Mount Everest that has yet to be repeated. Due to political obstruction, the next expedition didn't follow for seven years. When it did, the first complete traverse of the four 8,000-meter summits of Kangchenjunga—one of the great feats in Himalayan history—fell to the Soviet climbers. Later, more Soviet climbers began to climb beyond their borders, racking up Himalayan successes that continue today. But the number of climbers who made it abroad was nothing compared to those who climbed only at home.

In 1991, when the Soviet Union fell apart, Soviet climbers gained complete freedom to travel. But they lost much more. The alpine centers, the training camps and the professional teams disappeared. Today, a few commercial firms have stepped into the void, funding expeditions to the Himalayan peaks, and some climbers have organized themselves into private trade unions, paying money from their salaries into a common fund to keep climbing. But training for large numbers of new climbers is finished, and only the most ardent of the old school survive.

And yet the proving ground of the Soviet climbers remains, and is now open for the West—and its wallets—to explore. Soon, the walls of the Pamir-Alai will be famous, and the hospitality of its nomadic people will be widely known. Climbers from around the world will explore the Pamir, Tien Shan and Ala Archa, the Kokshaal-Tau, Ak-Tau and Karacol. With them will come some money. Other money, one hopes for the sake of the former republics, will come from other sources, and the Commonwealth of Independent States may find its economic heartbeat, struggle back to its feet, and begin to function once more. Then, maybe, its climbing will enjoy a renaissance—but, until then, there isn't enough money for the people to go to the mountains, and the system that raised generations of the world's best mountaineers is forever gone.

to die for

Coral Bowman
July 1998

"I'm not one to let go easily, which was a good thing in climbing, but not so good in my personal life," writes Coral Bowman of the central event in this riveting story. In a disastrous fall, Bowman escaped almost certain death through a miraculous act of self-preservation. Afterward, she could not escape the realization that climbing, and the rest of her life, would never be the same.

The summer I turned 13, they built an Olympic diving platform at Woods Pool. Almost every day, my friends and I rode our bikes to the pool and mustered up the courage to jump from the top. The first levels were easy. But the top took my breath away. It seemed impossibly high. The day I finally jumped, my arms and legs smacked the water so hard I cried. Although I jumped halfheartedly a few more times, just to show I wasn't afraid, it never got any better. But hitting the water at least stopped the far worse sensation of being alone and falling freely, unendingly, through space.

Ten years later, when I moved to Colorado, I discovered rock climbing, a sport that seemed to fit my body and psyche. It didn't take long before I knew how to climb safely and well. Very well. Then one autumn day I made a mistake that would change my life forever. To many, that I lived to tell about it was a miracle. I too was grateful to be alive. But my illusion that I would find love and happiness through my climbing was shattered. And the attention I'd desperately sought from climbing came to me, ironically, because of a fall. For a while, I was "the girl who fell off the Naked Edge."

■ ▪ ■

September 12 dawned clear and bright. It was the last day for Sue and I to try the Naked Edge before she left town. All summer, we'd been climbing with this goal in mind. We wanted to do the first all-female ascent of the Edge together, and we figured if we didn't do it before Sue left, someone else would beat us to it. Already, Beth had done the first female ascent with a young man from Germany.

It was a perfect Colorado morning, but also a lousy one. The love of my life for the past three years was supposed to be leaving for the Gunks in a few days. He would be gone for several months, and I had decided not to go with him. Neither climbing nor my relationship had brought any joy the past four months. I needed a break from climbing and, I decided, from him. Distressed by the unhappy state of my personal life, I had slept poorly on someone's front porch in Eldorado Springs. When daylight came, I groaned and anxiously readied myself to climb.

True courage would have been to tell Sue I was feeling too sad and sorry for myself to climb that day. But I didn't understand these things yet—I was a climbing machine. I'd trained myself to sublimate all feelings in pursuit of being on top. I functioned on automatic pilot. Worries, fears, doubts and anxieties had to be set aside in order to climb well, to climb hard. And, usually, it was a relief to leave my personal life at the bottom of a cliff and just focus on the rock beneath my hands and feet. I hadn't learned that some days it was better not to get on the rock.

So, with my best "do or die" attitude, I set off. We more or less third-classed up to the base of the Edge, trailing the rope between belays, but using little protection. Fretting on one such belay, wondering what was taking Sue so long, I heard an unmistakable voice calling my name. Looking down, I saw my on-again, off-again love beaming up at me. The last time I'd seen him he hadn't even wanted to look in my eyes, let alone be close to me. Now he was calling out, "Looks like it's going to be a good day."

Damn. How dare he be so friendly, so nonchalant, when we were hardly speaking? My heart ached every time I saw him. I wanted to go back to the simple days when all that mattered was climbing and being with him. Life off the rock just seemed too complicated.

"Good luck," he called out and turned with his partner in the direction of the West Ridge.

I felt the rope pull tight on my waist, and I called out, "Climbing," grateful to begin moving up the rock. Here I felt safe, in the quiet world of rope, rock and sky.

■ ■ ■

Because we were uncertain about the weather and how long it would take us to do the climb, we had packed sweaters and water bottles. Neither of us relished doing the hard pitches with a pack on our back, so I had brought a 9mm rope so we could haul the pack instead.

Although I would have liked to have done the first and hardest lead, I knew Sue wanted it too. And she definitely didn't want to lead the slot on the fourth pitch, so we decided she'd get the first, third and fifth pitches, and I'd do the second and fourth. I owed her one anyway, as she had frequently let me have the crux on other climbs.

From the belay, the first pitch looked deceptively short but thin and spidery. Sue messed around with protection, and I had to bite my lip not to hurry her. I knew her style—she would make sure she was happy with the protection before she would commit. But she spent so much time that her arms blew out and she had to grab a nut. Perhaps sensing my impatience, she said she would just use the piece as aid and finish the pitch.

I wasn't for this. I hated the delay, but wanted a clean free ascent, and I persuaded her to come down and rest. I knew she could do it, and the second time she made it look easy. How much time had passed I couldn't tell, but some clouds had appeared, and I was impatient. I followed up the first pitch hurriedly, until it got hard enough that I had to take it seriously. Stemming madly to the left, I moved through the crux, impressed by the lead. It had been harder than it looked from the ledge, and I was glad to follow it.

The second pitch seemed pretty inconsequential. It turned the arête, which was spectacular, and I was up it quickly. Sue joined me on a small sloping belay, and we hauled up the pack. The wind had picked up, and we debated what to do with the ropes. I decided to drape the red climbing rope loosely over my foot, to keep control of it, and to let the orange haul rope blow in the wind. As Sue led off out of sight on the third pitch, the orange rope billowed out around the arête and disappeared.

Already, I was wanting this climb and day to be over. Maybe this week as well, when my two closest mates would be leaving town for months. Sue's voice brought me back from my dismal reverie. She was getting a lot of drag from the haul rope and couldn't climb.

I tied off her belay rope and struggled with the orange 9mm. It wouldn't budge. Annoyed at myself for not managing the rope better, I yelled up to Sue that I would belay her back down and then rappel down myself to free the rope below. Although the roar of wind and water made communication difficult, I felt Sue's rope begin to slide down toward me, and soon she was beside me again.

Inside, my frustration was growing. It seemed like hours had passed, and we still had three more pitches to go. Three more pitches I didn't want to climb. In a flurry, I took charge. Sue clipped into the anchor and listened to my plan. I would rappel down on the red rope and free the orange haul rope, then she could belay me back up the second pitch. We both untied from the red climbing rope, and I clipped it through a sling with two carabiners. Then I set up my brake system. I always used two carabiners with the gates reversed and two across them for the brake.

Sue barely had time to think before I was backing down the sloping rock below us.

At the lip of a small overhang, I leaned way back so I wouldn't swing against the rock as I cleared the roof. Then I gave a small push off, anticipating that giddy feeling of sliding down the rope before I touched rock again. But this time was different. I heard a strange, metallic click and began to drop. My eyes met Sue's and I mouthed, "What the f—?" Then I was falling fast. I was alone and falling freely, unendingly, through space, and there was no pool below.

Time seemed to slow, then stop. I turned, looking over my shoulder at the ground, several hundred feet below. In a detached way, I knew that, when I hit it, I would die. As I began to turn back to the rock, a vision of my friends appeared, a huge crowd of people watching me, smiling, concerned. "There are many people who love me and care about me," I thought. Then I heard myself say fiercely, "I don't want to go. I'm not ready to die yet."

My mind snapped back and there was the orange haul rope in front of me. That damned rope I'd been cursing moments before. I'd probably fallen 20 feet by now and must have been moving very fast. I reached out and grabbed on, thinking, "Thank God, I'm safe." But my hands were on fire as I slid down the thin rope. "I'll never be able to hold on," I thought. "It hurts too much. I'm going to die after all."

But as quickly as the fall began, it miraculously stopped. The hand of God had held mine, and I gave a sigh of relief to be alive. Stretching my legs out in front of me, I could just touch the rock face. Although I had stopped my fall with my bare hands, my difficulties were not over. Now I relied on my will and adrenaline to survive.

Looking at my brake system, I was surprised to see the red 11mm rope dangling uselessly from it. My brake hadn't failed—the rope had come unhooked from the anchor. There was no time to ponder how, for I was connected to the haul rope only by my two small hands.

Somehow they no longer hurt, and foolishly I thought, "Maybe they aren't burned after all." But as I let go with my right hand to wrap the orange rope around my right leg, I saw white marks oozing across my fingers where the rope had burned me. I needed to act swiftly while my hands could still function.

Looping tangles of the useless red rope around my other leg, I clipped a loop of it to my gear sling. I hung on tightly with my left hand, and managed to unclip the cross carabiners of my brake system, freeing the red rope. Quickly, I pushed a single strand of the orange 9mm up through the carabiners and clipped two across. I had no more biners to put on.

This would be a fast descent, and my hands were rapidly becoming stiff and useless. I left the orange rope wrapped around my right leg several times to try and control my slide. Then I swung in to the rock, grabbing with my finger tips and pulling myself around the arête. Finally, I lowered myself to the belay at the top of the first pitch of the Edge.

Slowly and awkwardly, I clipped into the fixed anchor. The awfulness of what had happened sank in. I was alive. I was safe even. But I had almost died. And my precious hands were badly burned. "Sue," I called out in what seemed like a whisper. "Help me."

Feeling frail and helpless, I slumped against the rock. I had no idea how long it took her to get to me; I was too stunned and drained. Then she was just above me, rappelling down the orange haul rope. Though its mismanagement had been my downfall, it was saving us in the end. When our eyes met, I knew we were feeling the same thing: Thank God I was still here. While she secured herself, I asked what had happened.

It had been a freak accident. I'd used double carabiners to set up the rappel anchor, but in my impatience I hadn't reversed the gates. Both of us were careful, conscientious climbers, and we trusted each other's judgment, so Sue hadn't checked me. When I leaned out at the lip of the roof, the sling must have crossed the carabiner gates and pushed them open, and the rappel rope had popped out.

Now I was quiet and Sue took over. My hands were no longer functional, so after sorting out the red 11mm rope, she lowered me down the first pitch of the Edge. After her hard-won victory on this pitch, I was painfully sorry for her that we were ending the climb this way, in defeat. But we were both still too concerned about getting me down to talk much. Waiting for her to arrive, I gazed out across an empty canyon. I was relieved no one had seen me fall.

When Sue joined me, she spelled out two options: Wait for someone to come up and bring another rope to rappel from that point, or belay me down the sloping ledge to the rappel point by the Whale's Tail. Our single rope would reach the ground from there, and she could lower me again.

Not wanting the humiliation of a rescue, I assured her that I could manage. Using only my elbows for balance, I quickly traversed down to the anchor bolt. By now, my hands had swollen shut, so I couldn't even clip myself in. I sat numbly, awash in the agony of what I had brought on myself.

"Rock," someone yelled far above me. Shaking myself out of my stupor, I looked up to see a fist-sized rock hurtling toward me. I dove sideways and the rock smashed right next to the anchor, where I should have been clipped in and sitting. "God," I prayed, "please get me off this rock alive."

Sue lowered me down the last pitch to the ground. The earth had never felt so good. We walked out of the canyon and asked someone to take me to the hospital. I remember nothing of that trip.

■ ■ ■

In the aftermath, my hands healed pretty quickly, but my heart did not. My boyfriend went east three days later. He called a few times, missing me now that I wasn't around. But something inside of me had died. That lust for power, glory or love at any cost had withered. Somehow, I had to begin putting my life back together.

I got back on the rock three weeks later. Climbing was still my life. But my focus began to change, and I never did really hard climbs again. Eventually, I started a climbing school called Great Herizons, which brought a sense of joy and satisfaction to my climbing that had eluded me in earlier days.

It took more than three years to finally end my relationship and move on to other loves. I'm not one to let go easily, which was a good thing in climbing, but not so helpful in my personal life. I did learn something from this experience, however. No climb or person is "to die for." And I've never come so close again.

# of odds and angels

Andrew Todhunter
September 1998

Andrew Todhunter wrote an insightful book about Dan Osman, a climber who took to staging huge falls onto climbing ropes and eventually developed the skill and nerve to jump over 1,000 feet. Less than a year after the book's publication, Osman was killed in such a jump in Yosemite. In this short excerpt, published in Rock & Ice *a few months before Osman's death, Todhunter examines the risks he himself has taken and accepts the inevitable slow-down as he begins to raise a family.*

At this time in my life, a watershed of sorts, I am unwilling to ride a motorcycle on the street. I so abused the generosity of the Fates that I now fear them as a penniless gambler fears his creditors. I have outstanding debts in this regard, and, with a child on the way, I cannot afford to pay them. One of my many fears as an expectant father is that a child of mine will one day appear in the driveway on the back of a motorcycle.

Until recently, I never realized to what degree the middle class is a moral, rather than an economic, entity, based less on status than on possessive, life-sustaining love. The moral middle class plays life by percentages, because that is the safest bet, and percentage play—in life as in tennis—is by definition conservative. It is, ironically, the gambler's game. For all appearances, a true gambler never makes a poor bet. He never goes for the long shot. If he is a professional, he always bets the sure thing, and, even then, he occasionally loses. When the currency in question is your family's welfare, there appears to be no other way.

There is a deepening to this kind of life, informed by what the Japanese call *yugen*. An emotion akin to nostalgia, yugen might be experienced—as philosopher Alan Watts put it—in an empty banquet hall the day after a wedding. Napkins lie beneath the chairs, the rice crunches underfoot, but the music and the bride and groom and all the guests are gone. The Japanese compare yugen to the sight of a ship's sails slowly disappearing around a distant headland, or of geese, in formation, flying into mist. Living a life based essentially upon the protection and preservation of your family is to pit yourself against an enemy—the nature of all beings and all things to change and vanish—that you cannot possibly defeat. This is a rich conundrum, offering its own rewards. But it's not nearly as much fun, I'm discovering, as riding a Kawasaki GPz at 140 miles per hour across the Bay Bridge at three o'clock in the morning.

I still consider buying a bike for the track, a high-performance, street-illegal machine with no license plates and fat, slick tires as tacky in the heat as licorice. I would drive it to the track on a trailer and pay by the hour to ride. A track is blessedly free of obstructions, and the chance of serious injury or death after a fall is infinitesimal. In full leathers and a new helmet, on a clean, beveled track—free of oil, sand, black ice, potholes, mufflers, telephone poles, reckless pedestrians and desperate, clinging fathers in Volvo station wagons—I could still allow myself to wind it out. And yet, without the terrible, ambient risk, I wonder, would the enormous speed and the rhythm of the corners all too soon become monotonous? As I slowed and left the track and rolled the bike onto the trailer, while I lashed it into place for the drive home, to my wife, to my future child, would I feel as I've begun to feel now in the company of accomplished climbers? That I have changed, that another calling—that of family—has waylaid me, that I am equal to its tasks, that I perform my duty, but that I have chewed through my own entrails to do so, that I am no longer, in some essential way, entirely alive?

This is a paltry price to pay, I answer, for a wife beyond all expectation, and for the child she carries—a nameless, genderless child who illuminates the way before us like a brazier. Eventually, perhaps, these vestigial adolescent twinges will subside. I would happily trade everything I had at 20 for my life now, at 30. Like the lobsters I used to drag from caves on the floor of the Pacific—wise, ageless creatures that fought and fought until they finally surrendered, all at once, and went limp in the hand—I am molting, I hope, into a better form.

Early in my wife's pregnancy, I was toproping a 5.10 in a climbing gym in the adjacent town of San Rafael. I reached the gentle overhang and was prepared to move through it—I had climbed the route successfully before—when I unexpectedly froze in place. I thought of the baby. The music, cycled perpetually through the gym's loudspeakers, overwhelmed me. I glanced down at my belayer. Her belay appeared sound, but she was chatting with her neighbor, her eyes averted from the route. My motivation to ascend evaporated. The strength drained from my arms. My position—clutching like a salamander to misshapen plastic knobs bolted to a sheet of painted, artificially textured plywood—struck me as absurd. My fear unraveled quickly into panic. I lost all faith in the anchor, the rope, my harness, my belayer, and my ability to climb; I wanted down. I called to my belayer, trying to conceal my alarm, to watch me. Rather than rely on the gear, I decided to attempt the downclimb. Clinging to each hold, I descended in horrible, contracted form to the gravel floor.

At the base of the route, I untied from the rope, mumbled something to my belayer and excused myself. I sat in my car in the parking lot, in the darkness, and wondered if I would ever climb again. I unwound the strips of white athletic tape from my fingers, and rubbed the streaks of residual glue from my ring finger. From my wallet, tucked above the visor, I extracted and replaced my wedding ring. I studied the band of white gold, plain as silver in the light of the streetlamps, and drove slowly home.

For practical reasons, I no longer climb wearing the ring. It was a difficult decision, the first time I went bouldering after the wedding. I considered taping over it, to protect the metal, but this posed its own problems. Glue would foul the ring, and the resulting lump would be even more awkward than the ring itself. At first, I wore the ring on a chain around my neck, tucked under my shirt. But I soon became afraid of the chain breaking, and of losing the ring forever. Thereafter I strung the ring on my leather watchband and stashed it deep in a pack pocket, or tucked it into the farthest corner of my wallet. Taking it off and putting it back on became a minor ritual. I would study the appearance of my hand in contrast. With ring; without ring. Removing it, I continue to feel a faint pang of risk, as if I am colluding with chaos; seeing it again in its place is as satisfying to the eye as a tarnished copper kettle, its brilliance surfacing beneath the labor of a polishing hand. And yet the ring—or what it represents—is heavy; it gives weight.

A week of inactivity and self-loathing passed before I began climbing again, this time with greater caution. I drove with circumspection. For the first time, I pondered life insurance. I took up tennis. In the San Rafael gym, as I clung to that overhang and imagined the baby, then small enough to fit in my closed palm, I felt painfully exposed. In the minute, infinitely fragile fetus, I perceived the frailty of things—myself included—in a way I never had before. Furthermore, I suspected with the confirmation of this life in utero, my personal exemption from disaster had expired. My luck was up, and the angels had diverted their attentions to the child. They had preserved me through my youth for this transmission; through conception I had passed them on. Get down, said a voice. Get down. There is danger here, and it does not serve.

about the contributors

PAT AMENT has produced books and essays in the mountaineering genre for thirty-six years. He is a poet, artist, songwriter, pianist, karate black belt, photographer, and chess expert. He is married, with two children.

JAY ANDERSON and his word processor live in a cabin in the Sierras. He keeps current with the newest offwidths and is looking for the next Lucille.

MARTIN ATKINSON was a prolific climber in the 80s in the United Kingdom, France, and the United States, and is credited with the first ascent of many routes, including "Mecca," one of the hardest sport routes in the United Kingdom. He is currently managing director of Wild Country Ltd., and lives in Derbyshire with his wife and three small daughters.

MIKE BEARZI made the first free ascent of Cerro Torre in 1986 via the Ferrari West Face Route with Eric Winkelman. Since Andromeda Strain, he has attempted a new route on Cerro Torre and a couple of oxygenless alpine-style ascents of Everest. He lives in Colorado, where he makes his living as a carpenter and contractor.

CHRISTIAN BECKWITH is editor of the *American Alpine Journal* and founder of the *Mountain Yodel*.

BARRY BLANCHARD is one of North America's foremost alpinists. An internationally certified mountain guide, he has climbed and led climbs in the Himalaya, Andes, Alps, and Northern Ranges. He and his wife, Catherine, live in Canmore, Alberta.

ABOUT THE CONTRIBUTORS

CORAL BOWMAN retired from climbing and instructing in 1985. She is currently a graduate student in journalism at the University of Colorado, Boulder. Although she has passionate memories of her climbing days, she now aspires to be a 5.14 writer.

JIM BRIDWELL is legendary for his climbs, which include the first one-day ascent of the Nose on El Capitan and the first alpine-style ascent of Cerro Torre.

JOHN BURBIDGE is an editor for Falcon Books in Missoula, Montana.

CAMERON M. BURNS is a climber, journalist, photographer, and author of several climbing guidebooks. He lives in Basalt, Colorado.

GREG CHILD has climbed in the Himalaya with thirteen expeditions, completing the first ascent of Shipton Spire and summitting K2, Everest, Gasherbrum IV, and the Trango Tower. The author of three collections of essays and winner of a Banff Mountain Book Festival Award for *Postcards from the Ledge*, he is recognized as one of the world's finest climbing writers.

WILL GADD, former editor at both *Rock & Ice* and *Climbing* magazine, writes about climbing and other sports. He is an active climber, paraglider, and kayaker. He holds the world record in distance for paragliding, as well as three gold medals from the Winter X Games for ice climbing.

JOE JOSEPHSON has dedicated eight years and three Subarus to exploring the Ghost River in the winter and summer. Since 1986, he has ice climbed all around North America, including Alaska, Quebec, Newfoundland, Colorado, and New England. He is known for first ascents of Sea of Vapors and Acid Howl, a solo ascent of Grand Central Couloir, and, with Steve House, establishing Call of the Wild, a new 7,500-foot alpine ice route on King Peak in the Yukon. He lives in Bozeman, Montana.

JEFF LONG, an accomplished climber, is the author of several novels and works of nonfiction. He lives in Boulder, Colorado.

JOHN LONG authored the *How to Rock Climb* series of climbing guides, as well as many essays and adventure stories. He is an influential rock climber and boulderer who also took up jungle exploration in the late 70s. He lives in Venice, California.

MICHAEL G. LOSO lives in McCarthy, Alaska, where he serves as director of the Wrangell Mountains Center, a nonprofit environmental education and research organization. He has been an assistant guide, mountaineering ranger, and rescue volunteer on Denali.

JOSH LOWELL is a climber, writer, and filmmaker from New York. His company, Big Up Productions, has produced three climbing films: *Big Up: Bouldering in the Gunks, Free Hueco,* and *Rampage: On Tour with Chris Sharma.*

DOUGALD MACDONALD is the editor and publisher of *Rock & Ice*. He has been climbing for more than twenty years on four continents.

ALISON OSIUS is president of the American Alpine Club, senior editor of *Climbing* magazine, and author of *Second Ascent: The Story of Hugh Herr.* She is a former national champion and world-event competitor in sport climbing. She lives in Carbondale, Colorado, with her husband, Michael Benge, and two young sons.

NICK PAPA has been a climbing addict since his early twenties. Periodic bouts with the job world interrupt his extended climbing road trips. Fred retired to Primarily Primates in Texas, where he swings with a Chinese macaque and vies for Alpha status.

ERIC PERLMAN, co-director, co-producer, and co-cinematographer of the *Masters of Stone* video series, has been climbing in Yosemite for more than twenty years. He is a writer, photographer, cinematographer, climber, and ski mountaineer based in Truckee, California.

RUARIDH PRINGLE is an outdoor writer and photographer who lives near Edinburgh, Scotland.

ANDREW TODHUNTER is an amateur climber and adventurer who writes about extreme sports. He lives in Paris with his wife and young daughter.

MARK TWIGHT is one of America's leading alpinists. His stories and photographs of extreme climbing have appeared in climbing magazines around the world. Most recently, he is the author of *Extreme Alpinism: Climbing Light, Fast, and High* (The Mountaineers Books).

JONATHAN WATERMAN has written and edited seven books. He is an alpinist, adventurer, writer, and photographer.

ROCK&ICE

Rock & Ice is published eight times a year in Boulder, Colorado. For subscription information, call 877-ROCKICE (toll-free) or 303-499-8410 outside the U.S. and Canada.

THE MOUNTAINEERS, founded in 1906, is a nonprofit outdoor activity and conservation club, whose mission is "to explore, study, preserve, and enjoy the natural beauty of the outdoors.... " Based in Seattle, Washington, the club is now the third-largest such organization in the United States, with 15,000 members and five branches throughout Washington State.

The Mountaineers sponsors both classes and year-round outdoor activities in the Pacific Northwest, which include hiking, mountain climbing, ski-touring, snowshoeing, bicycling, camping, kayaking and canoeing, nature study, sailing, and adventure travel. The club's conservation division supports environmental causes through educational activities, sponsoring legislation, and presenting informational programs. All club activities are led by skilled, experienced volunteers, who are dedicated to promoting safe and responsible enjoyment and preservation of the outdoors.

If you would like to participate in these organized outdoor activities or the club's programs, consider a membership in The Mountaineers. For information and an application, write or call The Mountaineers, Club Headquarters, 300 Third Avenue West, Seattle, Washington 98119; (206) 284-6310.

The Mountaineers Books, an active, nonprofit publishing program of the club, produces guidebooks, instructional texts, historical works, natural history guides, and works on environmental conservation. All books produced by The Mountaineers are aimed at fulfilling the club's mission.

Send or call for our catalog of more than 300 outdoor titles:

The Mountaineers Books
1001 SW Klickitat Way, Suite 201
Seattle, WA 98134
800-553-4453

mbooks@mountaineers.org
www.mountaineersbooks.org

Other titles you may enjoy from The Mountaineers:

MOUNTAINEERING: The Freedom of the Hills, 6th Edition, *The Mountaineers*
The completely revised and expanded edition of the best-selling mountaineering "how-to" book of all time—required reading for all climbers.

A LIFE ON THE EDGE: Memoirs of Everest and Beyond, *Jim Whittaker*
Whittaker, the first American to summit Everest, tells the story behind the many stunning successes of his career in this extraordinary memoir. CEO of Recreational Equipment, Inc.; confidante of Bobby Kennedy; leader of the 1990 International Peace Climb; explorer and sailor who is circumnavigating the world with his family: here is an American hero revealed.

THE BURGESS BOOK OF LIES, *Adrian & Alan Burgess*
The tall tales, edge-of-your-seat adventures and poignant stories from identical twins and accomplished mountaineers Adrian and Alan Burgess. Full of stories from major climbs all over the world, this is a great read for climbers and armchair adventurers alike.

SHERMAN EXPOSED: Slightly Censored Climbing Stories, *John Sherman*
A hilarious and irreverent collection including the best of Sherman's "Verm's World" columns for *Climbing* magazine, plus profiles of prominent climbers and previously unpublished essays.

DARK SHADOWS FALLING, *Joe Simpson*
Troubled by the 1996 events on Mount Everest, veteran mountaineer Simpson boldly speaks out on declining ethical standards in mountaineering.

POSTCARDS FROM THE LEDGE: Collected Mountaineering Writings of Greg Child, *Greg Child*
Sharp, incisive, and irreverent, this masterful storyteller entertains even as he plumbs the art and culture of the sport of mountaineering.

ERIC SHIPTON: Everest and Beyond, *Peter Steele*
Steele draws upon scores of personal interviews as well as Shipton's own correspondence to draw a complete portrait of the self-effacing explorer, with new information about his public and private life.

REINHOLD MESSNER, FREE SPIRIT: A Climber's Life, *Reinhold Messner*
One of history's greatest Himalayan mountaineers, Messner reveals the forces and events that have shaped him as an individual and as a climber in this classic biography.

ICE WORLD: Techniques and Experiences of Modern Ice Climbing, *Jeff Lowe*
Renowned climbing veteran Jeff Lowe shares personal stories and professional insight in this comprehensive book, which offers a history of the fascinating sport and an overview of the world's best ice climbs.